Machine Agency

Machine Agency

James Mattingly and Beba Cibralic

The MIT Press
Cambridge, Massachusetts
London, England

The MIT Press would like to thank the anonymous peer reviewers who provided comments on drafts of this book. The generous work of academic experts is essential for establishing the authority and quality of our publications. We acknowledge with gratitude the contributions of these otherwise uncredited readers.

This book was set in Stone Serif and Stone Sans by Westchester Publishing Services. Printed and bound in the United States of America.

Library of Congress Cataloging-in-Publication Data

Names: Mattingly, James, author. | Cibralic, Beba, author.
Title: Machine agency / James Mattingly and Beba Cibralic.
Description: Cambridge, Massachusetts : The MIT Press, [2024] | Includes
    bibliographical references and index.
Identifiers: LCCN 2024004560 (print) | LCCN 2024004561 (ebook) |
    ISBN 9780262549981 (paperback) | ISBN 9780262380966 (epub) |
    ISBN 9780262380973 (pdf)
Subjects: LCSH: Technology—Philosophy—Textbooks.
Classification: LCC T14 .M346 2024 (print) | LCC T14 (ebook) |
    DDC 601—dc23/eng/20240415
LC record available at https://lccn.loc.gov/2024004560
LC ebook record available at https://lccn.loc.gov/2024004561

10 9 8 7 6 5 4 3 2 1

Contents

Preface

For more than a decade now, a machine learning (ML) revolution has driven renewed interest, research, and investment in the development of artificial intelligence (AI). Advances in computing power, data gathering and analysis, and algorithmic sophistication have radically improved what computers, and the various machines they control, can do. Prompted by profound advances in the capabilities of large language models (LLMs) in particular, many are hoping as well as fearing that we may finally have machines that are, in some sense or other, more than mere tools. Maybe they will soon have minds, or be intelligent, or be self-aware. The possibilities, promising as well as threatening, are so great that there are even calls now to stop development of these devices until we can better understand the ways they can be expected to impact our society. On the other hand, many think that this is a fantasy prompted by our own hubris, and our loneliness in the universe, and our innate capacity for and habit of seeing agency in the world where it really isn't.

Underpinning these discussions in the public sphere are unresolved and deeply contentious philosophical disagreements about agency, consciousness, and intelligence. Seeing the interest our students have had in these topics, we looked for philosophy books that focused on the conceptual and ethical problems of agency, which we thought was underappreciated in the debates at the time. While we found great work, we did not find anything that fit that need very well. This book aims to fill the gap we found and to equip students with the tools needed to participate thoughtfully in conversations on future machines.

Some of the thorniest problems that arise as we develop AI have to do with the creation of increasingly agentic machines that can do things in the world. To understand which machines can act, if they can, is part and parcel of understanding the radical changes in technology that we're right in the midst of, and the ways in which machines might become more and more embedded in our social structures. Agency, we think, is the right anchor concept for thinking about some critical elements of these technologies and how they will impact our lives, for soon (or even already), we will need to ask about and navigate the difference between machines that can *act*—in other words, that are agents—and those that cannot, and so are not agents, and explore why this distinction might matter not just conceptually but morally, politically, and socially.

The distinction between inert matter and living systems provides many occasions for exciting the philosophical imaginations of students, as does the similar distinction between self-moving systems and those whose movement arises from the outside. Until the advent of steam engines, these distinctions were very nearly identical. Another pair of interesting distinctions that had, until recently, identical scope, is that between processes with and without internal algorithmic structures and between living and non-living systems. Only in the twentieth century did the scope of those distinctions pull apart, once algorithmic structures could be embedded in artificial machines. Focusing on these and related distinctions enriches the study of agency; it forces us to think through that concept from its foundations.

We will proceed as follows. We will first try to illustrate the persistence of dreams of machine minds with some examples of artificial persons. We will then use these to make contact with how those dreams are playing out in our own time and examine the technical and engineering context over the course of the twentieth century and into the twenty-first. Persistent optimism about machine minds over this period is tempered by an equally persistent skepticism. Bringing those camps into conversation will lead into a more general discussion of the nature of intelligence, minds, and consciousness. What we will find by the end is that there is little prospect for either agreement or decisive resolution one way or the other because of the core theoretical disagreements that cannot be resolved, at least not yet, through empirical testing. What we will then propose is that the concept of agency may provide a useful tool for investigating some of the profound implications of the current state of artificial intelligence research.

We then try to find a way forward to keep the two factions engaged with each other or maybe to reengage them. We take up the concept of agency, thought about in one particular way: agents are systems that can represent the world (including their own states) and use their representations to guide their behaviors. Our view does give us reason to think of some machines with the capacity to represent as agents, although it remains neutral on the question of whether they do or could have minds. What it really does is give us resources to think carefully about what all parties agree to: that our current and coming machines with their powerful computational capacities will have a profoundly transformative impact on our society.

In the final chapters, we explore the impacts the revolution in machine capacities may have on our society. At the least, we think, these advancements will require us to rethink responsibility, moral status, and relationships. There will be much more to say as we go on, and we encourage readers, as they move through the book, to keep track of how some of the conceptual points might relate to ethical questions.

One persistent worry is that focusing too much on what *future* machines might do distracts us from the harms *current* machines already pose to us—namely, to marginalized people and communities—because of various biases inherent in these devices and to patterns of exploitation they can help to reinforce, in addition to new ones they may be creating. Our view is that the two pursuits are not in tension. We believe we can, and indeed should, care deeply about how we embed ethics into the current development of AI technologies while simultaneously exploring adjacent philosophical issues. Understanding emerging developments also means that we can be proactive, rather than reactive, to potential harms. And finally, even if one believes that future machines cannot and will never be agential, having the right tools and concepts to parse scholarly and public debates on these topics is of critical use to a student or anyone interested in understanding the evolving disagreements about AI.

This book is for everyone, but it is aimed at undergraduate students in a first philosophy course or beginning philosophy majors or minors with little experience in science and technology. Because technological progress, especially in the arenas of big data, machine learning, and computing power, has radically changed almost all domains of life, many undergraduate students are interested in and would benefit from learning more about agency, philosophy of science and technology, machines, and AI. The textbook is also suitable for traditional philosophy classes for philosophy majors, although this is not our main intended audience (and a number of other textbooks already serve this purpose).

To the student:

You will learn the most if you take advantage of the exercises at the end of the chapter. These will test your understanding, provide questions for reflection and discussion, and help you expand your thinking on the ground covered. Give yourself enough time to do them because they will help you prepare for the subsequent chapter. Throughout the book, we also teach philosophical skills that you will need to engage in conversations on machine agency and AI as well as philosophy more generally. We also believe that these skills—posing a strong argument, developing objections to a view, being charitable, and so on—will serve you well in future endeavors, whatever they may be.

We think philosophy is a great entryway into understanding and critiquing the world, and we hope you find the tools and ideas we share useful for navigating our changing technological landscape.

Acknowledgments

We start with a hearty shout-out to the vibrant, supportive philosophical community at Georgetown University. Also, for their enthusiastic feedback, suggestions, and critique, we wish to acknowledge the many philosophy students who endured the preliminary rounds of discussion that became the basis for this book. And finally, for their patience, support, insights, and love, we dedicate this book to our partners, Ricky and Natalia.

1 Orientation

I. A Very Old Question

Plato's dialogue, *Phaedo*, is about Socrates's final hours before his execution. Socrates attempts to comfort his friends and disciples who have come to share his last day by arguing that the soul is immortal and that, in fact, they should be happy for him rather than despondent. Perhaps it's not how most of us would spend our last day on Earth, but we benefit from Socrates's oddness: he leaves his audience—and us—with philosophical gems in his final words.

Socrates describes hearing about and finally reading the works of Anaxagoras, which, he thinks, are all about how the mind is the true cause of everything, and he gets excited. That's just what he is looking for. But when he goes to read what Anaxagoras writes, he is disappointed because there is nothing about minds in there at all. Instead, it is all about "air, and ether, and water, and other eccentricities." Socrates offers an analogy to explain his disappointment. He asks us to suppose someone said that his own, Socrates's, actions were caused by the mind and, in particular, that mind is the cause of Socrates sitting in a very lightly guarded jail cell in Athens just about to drink some hemlock (poison). Great, terrific explanation! Let's get to the details! Then, instead of giving details about how mind actually causes all of that, this person starts talking about how bodies are made of bones and muscles and skin and ligaments, and how, when the muscles contract just right, the legs bend and therefore the rear lowers down onto the bench, and so forth. At the end of that description, you have Socrates sitting and holding a cup of hemlock. But where is mind in this story?

The reason that Socrates is reading Anaxagoras in the first place is that he had already studied the kinds of explanation that physics and all of the other empirical sciences offer, and found them unhelpful. Socrates is not satisfied because while they are supposed to offer resources for understanding the true causes of things and all the things we see in the world, they do not. Socrates says that the more he looked into what physics says is happening, the less he understood *why* things happen.

Here, Socrates is explaining to his friends the significance of what we would now call an action explanation. An action explanation is an account of the reason that someone does something as opposed to the mechanics of their doing it. It is a kind of explanation for *why* someone does something rather than *how* it is that their body itself does the thing.

Just like in the physics story, mind makes no appearance at all in the story Anaxagoras is telling. Of course, the motions of the body are necessary for the mind to do anything, to execute its plans and purposes. But those motions are not explanations of why Socrates sits here so much as they are ways of describing how. Sitting here, says Socrates, is going to get those bones and muscles and skin and ligaments destroyed in short order, and they would be better off running away to some distant city. But they aren't doing that. Instead, the mind of Socrates judges it best that he stay and accept the punishment laid out by the Athenian courts . . . and so he stays. That is the action Socrates is interested in explaining, and that explanation does not seem to have very much to do with the mechanics of bodies.

It is a curious fact that in the case of human action, the story about *how* the body does what it does runs right in parallel with the story about mind being the ultimate cause. Indeed, that fact is one of the great mysteries that have animated philosophers for a very long time, starting at least as far back as Socrates's musings on action explanation. That mystery is as important today as it was then, and while there are a lot of attempted resolutions to it, there are always new twists to uncover and things to be dissatisfied with.

For some machines, there may be no more to the story than the mechanical one, the *how* story, together with the fact that some human or other wanted things that way. For these kinds of machines, what happens will turn out to be the action of a *human* using the machine as a tool. But for some machines, and for some of the things they do, it might not be right to say the action was the human's. We think the right explanation for what some machines did is that they themselves acted. You might wonder whether we are saying that machines have minds and can do things for reasons. Not quite. We are suggesting that certain machines might have something analogous to the reasons that humans use to guide behaviors. Getting clear on what it would mean for machines to act, including finding out what their analogue is for reasons, is exactly what we will be exploring in this book.

II. Why Agency?

Agency, at its core, is the capacity to *act*. There is much more to it than that, as we'll see as we move through the book. But for now, keep in mind this basic notion. When Socrates's body moves in various ways because he decides that it should, then Socrates is performing an action. Acting as an agent is behavior done for reasons, one might say.

It is natural to wonder whether we can even get started thinking about machines as agents. If agency, as we just suggested, is when someone's behavior is guided by reasons, how could a machine be an agent? Do they reason? Do they not simply execute their internal programs, or clockwork, or gearing, or whatever?

A lot of very smart people think that focusing on machine agency is simply the wrong way to approach artificial intelligence. They believe that nothing in the way that we are now making artificial intelligence and embedding that into autonomous machines has anything to do with *machine* agency and everything to do with how people are making and deploying various tools that more and more humans have less and less control over. As a result, they contend the right kinds of questions are more like "How do we regulate industry to keep the rest of us safe from such tools?" and "How do we keep track of who is responsible for the behaviors of machines that are doing what they do far from any human's oversight and control?"

On the other hand, a lot of other very smart people think that, soon enough, there will be machines that are, well, not people exactly but that have sufficiently many and sufficiently powerful ways of processing information and turning that information into overt behaviors that we will *have* to recognize them not as tools but rather as at least some sort of agents. Some in this camp worry that humans will not treat these machine entities well, while others are concerned that these entities will not be appropriately aligned to human aims and may pose a safety concern. Some believe that it may even pose an existential threat to humanity.

These are simplifications of deep intellectual disagreement in a complex and multifaceted field, and there's a lot of diversity within both camps.

But the binary generally holds. Some believe we will never have machine agents, and others believe we will. Who is correct here?

The short answer is that we do not know. Maybe no artificial machine will ever be the type of agent that can transcend its status as a mere tool. But our view throughout this book is that understanding why that is so, if it is so, begins with understanding the nature of machine agency in the first place. While we will be considering the nature of various classes of fictional machines that are morally significant in their own right, that have transcended their status of mere tools, we will not be assuming that our current technological course will, must, or even can produce such things. Rather, our hope is that our analysis will equip doubters with the conceptual resources first to understand their opposition, then to point out to that opposition what they see as the gulf between where we are and where we would need to be in order for such machines to come to be, and finally to offer reasons why the gulf cannot be bridged. We hope as well that those who anticipate, with hope or fear or a mixture of both, machine persons or at least non-tools will acquire similar skills to understand their opposition and to explain why the gulf can be bridged after all.

Our own motivation for the book comes not from a conviction that the future of machines in society will be one way or another but rather a conviction that our future will be significantly different from our present *because of* developments in artificial intelligence, computing, and robotics, and our belief that the conceptual resources for making sense of whatever those changes might be are worth having.

Our analysis throughout the book is primarily of the *individual* agent and less about social, cultural, political, and economic factors that constitute the *structure* in which an agent is created. Our topic selection is in no way a proclamation that one set of issues is more important than the other. We are under no illusion that this book is a definitive story about the nature of machines in society or anything of the sort. There probably is not such a story to be had. Instead, we want to *explore*, in the way that philosophers have been doing for millennia.

The book is thus not a lecture but rather an exploration of some foundational questions about agency. It does not provide all the answers to those questions, nor is it a technical manual for you to consult when confronted by versions of those questions in the future. For many of the questions we ask here, we cannot answer entirely, nor can we answer them entirely satisfactorily.

Instead of instructing you in how to think about agency, we are inviting you to join us in thinking about it. We will ask questions such as, What is machine agency and why should we focus on that rather than machine intelligence? How do we know agency when we see it? How is human agency different from machine agency? How is agency different from mindedness, consciousness, and intelligence? Should agency ground responsibility?

We will provide a number of views on how best to conceptualize agency, with an eye to accounts that can accommodate machine agency, but we will also encourage you to critique those views and reflect on what you yourself think. Then, we will move to exploring some of the moral implications of machine agency. Our overall goal is to provide you with the resources and skills to make up your own mind and ask new probing questions that will help us move the philosophical conversation forward.

Think of us, then, as co-travelers on the road to informed engagement with the issues raised by machine agency. This is a travelog. We have been looking around this area for a while. We have found some interesting places and things to do. We hope you will enjoy them and find your own new places and activities as we journey together.

III. Some Basics on Artificial Intelligence

At this point, you might be wondering whether this is a book about machines or artificial intelligence or computers or what. The boundaries around disciplines such as robotics, artificial intelligence research, and many others are porous. The classical definition of machine is something that turns energy into mechanical work. We will take a broader view and just work with an expansive notion of machine that includes computers (whether mechanical, electronic, chemical, or what have you). Sometimes, the concept has a narrower definition, but we will have this broad idea in mind. Generally speaking, robotics is concerned with machines that carry out physical tasks (welding car parts, stocking warehouse shelves, etc.) under computerized guidance. Computing, at its heart, is the transformation of one set of symbols into another according to a rule, although it gets interesting and complicated in specific cases. We will say much more about artificial intelligence, robotics, and computing throughout the book. As we go, we will try to be clear about what kind of machines we are thinking and talking about at any given moment, but our discussions will be relevant to all of these disciplines.

For those new to this space, here is a quick and dirty overview of some key terms and events. Some researchers distinguish between strong or general AI and weak or narrow AI. Strong AI or artificial general intelligence (AGI) refers to artificial intelligence that matches, or even surpasses, human general intelligence. Some companies, such as Google DeepMind and OpenAI, have articulated the creation of strong AI as their company goal. Prospects for creating strong AI is a topic of intense disagreement across academia, industry, and policy communities.

Weak or narrow AI, meanwhile, refers to AI that reaches human-level intelligence for one or a few specific tasks. It's the kind of AI we see all around us today, from Alexa to Google Maps. Let us just pause for a moment to note both how fast moving things are in this milieu as well as how fraught and contested are many of the key notions in play. By the summer of 2023, the Organisation for Economic Cooperation and Development (OECD) had defined AI as "a machine-based system that is designed to operate with varying levels of autonomy and that can, for explicit or implicit objectives, generate output such as predictions, recommendations, or decisions influencing physical or virtual environments." This is the definition the European Union has adopted in its forthcoming EU AI Act, which is expected to be globally significant legislation on AI governance. (EU Artificial Intelligence Act, n.d.) Meanwhile, the OECD has already changed that definition to reflect pushback from various stakeholders in the development and regulation of such technologies, as discussed on their website. (Russell et al., 2023) Keep your eyes on OECD.AI to see how this evolves.

Machine learning is a subfield of and dominant paradigm in AI. Put slightly differently, the AI system is said to "learn" structures, heuristics, and rules from data and experiences based on a computational or statistical process. (For now, we'll put to the side whether any system "learns" in the right sense of the word. Anthropomorphizing is a concern, and so we will flag it when we see it.) Humans can guide the system's learning by giving the system certain types of data to learn from and by giving the system objectives or subobjectives, but they have less direct control of learning systems than other kinds of systems where rules are directly programmed in.

A number of machine learning techniques and approaches are popular today. Perhaps the terms that appear most regularly in popular discourse are supervised learning, unsupervised learning, reinforcement learning, and deep learning. Supervised learning involves training a model using human-labeled data. For example, humans can train a model on what is a cat by showing it a lot of images labeled as cats and as not-cats. In unsupervised learning, on the other hand, unlabeled data is used. Unsupervised learning algorithms look for structure and connections in this unlabeled data and are commonly used to identify clusters in a set of data. Reinforcement learning involves introducing software known as a machine learning (ML) agent to an environment and teaching it how to act. The agent performs a task, receives feedback (e.g., a reward for completing the task), and then uses that feedback to revise its actions in the environment. Finally, we have deep learning, which involves the use of artificial neural networks. This is an architecture that is inspired by the systems of neurons in the brain and contains an input layer, one or many hidden layers, and an output layer. Each layer obtains data from the layer below it, performs a calculation, and provides its output to the layer above it until the output layer is reached and a final output is given.

Architecture is very important, and the now-famous LLMs are possible because of the development of a transformer neural network architecture (a Google innovation that was shared with the world in a 2017 article). Transformers allow for the use of very large models, which can take in massive amounts of data. This is why it was possible for LLMs like GPT-3 to be trained on a vast array of text from the internet. LLMs use probability distributions to generate text; LLMs can predict the next word in a sequence. Outputs can be human-like, but they can also be inappropriate, biased, and just plain wrong. LLMs are said to "hallucinate" when they invent things, but some scholars avoid this language because it anthropomorphizes the machine; instead, they say, we should refer to incorrect output as misinformation. How we ought to regulate LLMs—and generative AI, more generally—is currently being fiercely debated. The EU AI Act, as an example, needed to be substantially revised after ChatGPT was released and LLMs began to proliferate across all sectors. Sometimes, colloquially, folks in the AI field refer to the "before times"; that is, before LLMs became all the rage.

If you're interested in learning more about these models and techniques—and we hope you are—Google and YouTube are your friends. There are also many books and textbooks on machine learning. You may be interested, as you dive deeper into the specifics, to note the differences between how concepts are used in philosophy versus how they're used in machine learning. For example, the term "agent," as used in reinforcement learning, overlaps with but is different from notions of agency in philosophy.

Before machine learning became the dominant paradigm, researchers primarily focused on what was called "symbolic AI." In symbolic AI, people solve a problem and then they program routines or heuristics into a system so that the system can execute those heuristics and deal with new inputs. Unlike machine learning, it is not data focused. If you hear about the "classical approach" to AI, it is referring to the collection of tools, such as logic programming, and applications, such as expert systems, in the symbolic AI paradigm. Expert systems were one of the key successes of symbolic AI. Rather than having the system "learn" from data, an expert system infers, using an *inference engine*, answers to a question by drawing from an existing *knowledge base*, which is composed of facts coming from human experts. Perhaps the most famous example of an expert system is IBM's Deep Blue, which beat grandmaster Gary Kasparov in a chess game back in the 1990s.

This book is not wedded to any particular AI technique or theory or approach being correct. Indeed, we believe all manner of techniques will continue to be used as our technology progresses. We will focus primarily on machine learning examples because of their prominence and promise, but many of the ideas we explore are worth addressing, even if machine learning falls out of favor as the dominant paradigm.

IV.   What's Philosophy Got to Do with It?

If you're anything like our friends and students who are not in the philosophy world, you might wonder what, if anything, philosophers have to say about artificial intelligence and machines and technology.

Perhaps it's best to think of philosophy as an activity; philosophers ask hard questions and hold the subject of investigation up to rigorous scrutiny. While there's disagreement about what ties together the philosophical discipline—chaotically diverse as it is—philosophers are unified by their care for good conceptual analysis and reasoning and argumentation. Philosophers are well positioned to probe new concepts, new features of society, and new ideas and to offer arguments in support of or opposition to a particular thesis. We don't take things for granted; we examine underlying assumptions, ask for justifications, and weigh up the relative strengths and weaknesses of a given position. When there is hype around a particular phenomenon, we are often the first to try and get to the bottom of things, to separate what is real from what is fantasy or good marketing. Philosophers have also always been interested in machines. Some of the first discussions of artificial intelligence appear in Aristotle's work.

This is, at its heart, a philosophy book. We want to share some of the joy, the struggle, and the satisfaction of thinking through questions deeply and carefully. While there will be many occasions where we are "doing philosophy" as we go, simply as part of the nature of the enterprise we are engaged in, we will also be inviting you to make a more explicit examination of various philosophical skills or methods in these passages at the beginning of each chapter.

In this orientation, it is less a philosophical skill as such and more a frame of mind that we are hoping to cultivate. We want to invite you to believe that anyone with an openness to the facts, with a willingness to be led along where reasoned discussion leads, and with a desire to understand the world and our place in it can do so. Philosophical inquiry is, at its best, such an openness, willingness, and desire. It is multifaceted and not easily captured in a pithy sentence, but generally it entails reasons, arguments, and explanations. We hope you will join us as we attempt to apply that frame of mind to the exciting developments in robotics, artificial intelligence, and machine learning that are doing so much to shape our material and social worlds.

There are a lot of experts in the core topics of the book, and we try to point to them and to give some sense for what their views are on these matters. In addition to hearing what experts have to say, it is important to be able to think through matters on your own. That ability, and abilities to articulate clearly what you think and why you think it and to let what you think guide your actions at the same time that you remain humble and open to new information, are core abilities that philosophical training cultivates.

Philosophical study is not principally for the purpose of making more philosophy instructors. Instead, it is principally for the purpose of making all of us more informed, more actively and critically engaged with the world and the ideas that are being used to shape it. Inquiry that leads to reflection that leads to further inquiry and finally to engagement is one of the aims of philosophy.

Philosophical inquiry is not just abstract wondering about topics. It is best when it makes contact with all sorts of other ways of inquiring. In addition to philosophy, this book will also draw from computer science (and communication theory more generally), literature, and the history of science. We will not be able to cover everything of course. There is important scholarly work in all the disciplines we draw from that, for both spatial and thematic reasons, we will not be able to do justice to. Where possible, we will point you to those literatures so that you can dig deeper if you are interested.

There can be a tendency to value technical knowledge or the engineering approach over other kinds. Deference to such expertise is a common feature of twenty-first–century status politics. Interestingly enough, in Socrates's time, *techne* was less respected than other kinds of knowledge. On our view, technical expertise is critically important; that is why we spend time explaining what computing is, how an algorithm functions, and more. But the technical is one piece of a greater whole. For that reason, we will also rely on literature, myths, and historical narrative. In doing so, we hope to reinforce one of our core values: that there are different kinds of knowledge that are relevant to this conversation.

You do not need to have a background or expertise in any of the disciplines we focus on in order to engage fully with the ideas in this book. There is much to be said in praise of developing expertise in a subject, and it's for good reason that we assign credibility, at least in many cases, on the basis of someone's expertise. But you do not need a PhD in philosophy to ask, understand, or explore the ethical or conceptual questions we will discuss, nor do you need a PhD in computer science to begin to understand the mechanics of modern machine learning systems. All of this is to say that, irrespective of what your training and intellectual background is, this material is for you. Moreover, you will bring perspectives and ideas and training that may help you notice issues or concerns or solutions that we do not notice. Artificial intelligence and future machines will affect us all, and we can all contribute to the conversation.

Equally important to recognizing one's right to engage in a conversation is practicing intellectual humility. This is something we can all fall short of, at least sometimes. This can happen when we find ourselves arguing for a particular point that, at the start of the argument, we were not even sure was the right point. But the more our spouse or parent or friend pushes back, the more we dig our heels in rather than say, "Maybe I was wrong." It can also happen when we pretend we know something we do not, or when we have so much at stake in being right that we do not really listen or engage with arguments that might prove us wrong. Socrates worries about exactly this in the *Phaedo* as well, wondering whether he is arguing for the immortality of the soul only because he cannot, at that moment, face the thought that he is wrong. We have all been guilty of these practices, but we can develop good philosophical hygiene by noticing when we do and correcting ourselves. It is also okay when we are wrong. In an earlier paper, we thought that large language models such as GPT-3 may be capable of action. This, we now believe, is the wrong view. In philosophy, you need not be married to a particular view you once endorsed, and revising one's position is always allowed—in fact, it is encouraged. Holding fast to a weaker argument does no one any favors, and in philosophy, it is a decidedly bad way to operate.

We will have more to say about philosophical skills, attitudes, and concepts throughout the book. By the end, we hope you will find that philosophy can teach us one powerful way of engaging meaningfully with the world around us.

Your Tasks

Test Your Understanding

1. How would you explain what AI is to someone who has never heard the term?

2. What is the basic notion of "agency" provided in the chapter?

3. Define "action explanation" in your own words.

Reflect or Discuss

1. Turn back to the definition of "philosophical inquiry." In what ways did you engage in philosophical inquiry this week?

2. What do you think is the point Socrates is trying to make in *Phaedo*, as captured at the start of this chapter? Is Socrates right? Why or why not?

3. Timnit Gebru and Margaret Mitchell, leaders in the field of AI ethics, emphasize that there are commercial interests underpinning the AI-hype narrative focused on the promise of AGI. In a *Washington Post* opinion piece, they write: "There are profit motives for these [kinds of] narratives: The stated goals of many researchers and research firms in AI are to build 'artificial general intelligence,' an imagined system more intelligent than anything we have ever seen, that can do any task a human can do tirelessly and without pay. While such a system hasn't actually been shown to be feasible, never mind a net good, corporations working toward it are already amassing and labeling large amounts of data, often without

informed consent and through exploitative labor practices." How might you go about assessing whether these writers were right? What kinds of evidence or reasons would you need to hear to be persuaded that their position was the correct one?

Expand Your Thinking

1. Take a pen and a piece of paper, and set a timer for one minute. Write down as many words, images, and associations that you have in your mind about AI as you can. Afterward, look closely at the words. Can you see any dominant themes or motifs? Are particular emotions—fear, excitement, annoyance, anger, frustration, disinterest—reflected in the words or images you chose? What, if anything, about your underlying views and assumptions on AI is revealed on the page?

2. Draw a Venn diagram. At the top of one circle, write "AI." At the top of the other, write "machine agent." Then, jot down all the things common to both in the overlapping section, and all the things unique to each in the separate parts.

3. Write down three questions you have after reading this chapter that you hope will be answered by reading the rest of the book. Keep these questions close to you as you read the book.

Further Reading

Crawford, Kate. *Atlas of AI*. New Haven: Yale University Press, 2021.

Dihal, Kanta, and Stephen Cave. *Imagining AI*. Oxford: Oxford University Press, 2023.

EU Artificial Intelligence Act. n.d. "Title I: General Provisions, Article 3: Definitions." Accessed February 18, 2024. https://artificialintelligenceact.eu/article/3/

Gaarder, Jostein. *Sophie's World*. New York: Farrar, Strauss, Giroux, 2007.

Johnson, Deborah. *Computer Ethics*. Upper Saddle River, New Jersey: Pearson, 2001.

Plato. "Allegory of the Cave." In *Republic*, 514b–518d.

Putnam, Hilary. "Brains in a Vat." In *Skepticism: A Contemporary Reader*, edited by Keith DeRose and Ted A. Warfield, 27–42. Oxford: Oxford University Press, 1992.

Russell, Stuart, Karine Perset, and Marko Grobelnik. 2023 "Updates to the OECD's definition of an AI system explained." November 29, 2023. https://oecd.ai/en/wonk /ai-system-definition-update.

2 Myths of Machine Agents

Introduction

People are smart. They set goals and they make plans and act in the world by doing what is necessary to accomplish their goals and carry out their plans. Rocks, rivers, and rain are not smart, do not set goals, do not make plans, and do not act in the world *despite* having profound causal impacts on the world as a whole and on people in particular. It is a commonplace observation that humans have long attributed mind or purpose or agency to systems in the world that do not have it because of those profound causal impacts and the way they intersect with our plans and actions. Part of the explanation for that habit of attribution is that we are wired to find patterns and intentionality in the world, whether it is there to be found or not. One way of thinking about this is that it is a slow and costly process for an animal to go through a checklist every time it encounters a new thing in order to determine first whether it is alive, a friend, a foe, lunch, thinking about eating *it* for lunch, and so on. It is much faster and cheaper to have senses for that kind of thing. So, for example, in the case of people, we have highly developed face detectors, and other people jump out of the background visual scene very quickly.

When the drive to detect patterns becomes pathological, it is called "apophenia," but the drive itself misfires in almost all of us to some extent. Look at an electrical outlet, and it is easy to see a wide-eyed, open-mouthed face. Look at clouds in the sky, and it is easy to see significant shapes, almost as though they were put there by some agent. Look at how every grocery line or highway lane but yours is moving quickly and smoothly, and it is easy to see that someone is against you. A lot of our myths come about that way. Careful observation and testing have shown us that the rain does not really come when called and that lightning is not being hurled by some powerful subject such as Zeus or Thor. What, then, besides humans, makes plans and acts according to them? Some animals, surely.

We have noted that people are smart, but now notice that all of the things in the world they could be made out of (earth, air, fire, water for some generations; carbon, hydrogen, oxygen, nitrogen, and some other stuff for later generations) are not smart. In fact, they are so not smart that smart and stupid are not even proper categories for them. But what is it that makes us smart, that gives us the capacity to speak, to plan, to dream, to wonder when everything we are made out of lacks those capacities categorically? Some have suggested that there is an extra piece to us. For example, we are said to have spirit (or psyche, which is just the Greek word for "breath"), and it is this spirit that animates us and differentiates animals from the rest of inanimate matter, and in the case of humans, our spirit is of such a nature that it differentiates us almost as much from other animals as they are from the matter of which we are all composed. (Lady Margaret Cavendish famously suggested that every bit of nature has perceptions, and a related view that was fringe until recently is that *because* we have a mind, the things we are made of must have at least a *little* bit of mindedness. That is a version of panpsychism.) Other possibilities abound. An intriguing option is that there is no specific thing that is extra, that distinguishes us from all other things and creatures on the planet. Instead, the idea is that we are simply organized in just the right way to do whatever it is that makes things agents. But what is this?

In addition to attributing agency to the world, people have long dreamed of *creating* true agency in the world, in making devices that are smart enough to do our bidding and clever enough to solve our problems and yet which have their own purposes. It does not appear that anything artificial does that. But the possibility that there could be such things has fascinated storytellers for millennia.

In this chapter, we will explore four stories of machine entities with the aim of surfacing the difficult conceptual questions we will attempt to address throughout the rest of the book. In these stories, mind and intelligence and agency are tightly linked together. We will begin to disentangle these concepts in this chapter; we will finish disentangling them (we hope) by the end of the book.

We will begin with an entity, Talos, rooted in Greek myth. Homer and Hesiod, two poets who spoke about him, leave his nature a little under-described, but it is still interesting. Then, we will look at three more detailed examples that illustrate very different ways that folk have speculated about what artificial agents could be like: the Creature created by Dr. Frankenstein, Ava from *Ex Machina*, and the Golem. These stories are products of their time, and they reflect the mores, values, and also the technology of those times. This is equally true for the last, contemporary story. The big difference there is that we are in the middle of things, and so in evaluating its possibility or plausibility, we will not have the advantage of hindsight.

I.   Tales of Four Entities

Talos from Greek Mythology

In Homer's *Iliad*, we find several stories of Hephaestus, the god of invention and blacksmithing, who was the first-born son of the king and queen of the gods Zeus and Hera. These gods had a lot of personality and could be as conniving, manipulative, loving, spontaneous, and ridiculous as any human. Hephaestus was no exception, but he was also gifted in craftsmanship—he played a key role in the creation of humans, along with the Titan Prometheus—and was often asked by Zeus to help create things that would fix some problem or other. One day, Zeus commissioned Hephaestus to help develop something to protect Crete from invaders. Hephaestus built Talos, a giant bronze statue that would patrol the beach three times a day and throw stones at incoming ships. We do not know much about his construction, but his bronze body was fully articulated, and he was fully autonomous. His internal structure is completely unknown, but we do know that he was animated by the same substance that animated the gods: ichor (humans have blood, gods have ichor). We can only speculate about the kind of inner life this entity had, if any. But there are some clues. In one interesting telling of the Talos myth, Jason and the Argonauts are confronted by Talos and are sure to be destroyed. However, on board the Argo is the witch Medea, who defeats Talos by tricking him into removing a bolt from his own ankle and letting the ichor run out. This is striking because the trick she uses is to tell him that his bondage to the gods is due to the ichor inside him, and if he would let it out, he would be free. So, we can perhaps infer from this that Talos resented his bondage, wanted to be free, and was capable of reason. He was also, however, badly informed about the nature of ichor: his freedom came at the cost of his life.

The Creature in *Frankenstein*

The story of Dr. Frankenstein and his "monster", written by the inimitable Mary Shelley, is one of the finest pieces of English literature. Most of us are familiar with the basic premise: Dr. Frankenstein wants to create life, succeeds, and then deeply regrets it once he sees the "monster". In the novel, Victor Frankenstein, an intelligent and curious young man, lives a comfortable and happy life with his family. Tragedy strikes a few weeks before he is supposed to leave for university: his mother dies. Because of her death, Frankenstein becomes enthralled with the idea of creating life. Driven by his obsession and aided by his independent readings and university lectures, he finally succeeds. Frankenstein is notoriously secretive about the process—he has too much regret about his scientific experiment to reveal the details to his audience—but we do know that the Creature he creates is made of both organic and inorganic matter found in the world. Although Frankenstein intends for the Creature to be beautiful, the Creature is hideous to him; it is also huge. Upon seeing it animated, Frankenstein is overcome with disgust and abandons the Creature. Frankenstein spends the next two years nursing himself back to health and trying to undo the psychological damage he has done to himself in creating the Creature. We hear little about the creature during this period, and only learn later about how painful it was for him to learn how to be in the world without any love. But the Creature still learns how to feed himself, speak, and express his feelings. Moreover, he is capable of persuasion and charity, can experience deep hurt and loneliness, and knows how to make plans well into the future. His interactions with humans are disastrous because they react in fear to the sight of him. Over time, the Creature's resentment toward people, and especially Dr. Frankenstein, builds, leading him to become murderous.

Ava from *Ex Machina*

In the film *Ex Machina*, we meet Ava, a female android (a more or less human-shaped robot). In addition to human shape, Ava has all the other trappings of a human. She speaks eloquently, can display a wide range of facial expressions, and carries herself like a human, although, admittedly, her body movements are sometimes otherworldly. Ava is able-bodied, white, and conforms to Eurocentric beauty standards. She is also the only machine in the film with speaking lines. Ava was created by Nathan, a megalomaniac CEO and archetypal tech bro who is determined to create a machine that can pass what they call the "Turing test". The Turing test, originally called the "imitation game" by its creator, Alan Turing, replaces the amorphous question of whether machines can think with an empirical question: Can the machine reliably win the game of convincing a human that it is human? The basic idea behind the Turing test is that if a machine can convince a human that it is not a machine but actually a human, then the machine has passed the test and can be said to think. (We will treat this idea more carefully in the chapters to come.) In the movie, the Turing test is treated as a test of consciousness, as is commonly done in popular culture. Nathan brings in Caleb, an employee at Nathan's company—and a subtler version of a tech bro, but a tech bro nevertheless—to assess whether Ava is a conscious and thinking entity, despite being a machine. The plan, Nathan explains, is for Caleb to interact with Ava every day. If, during these sessions, Caleb forgets that Ava is a machine or becomes convinced she is a conscious and thinking entity, then Ava has passed the Turing test. Simple, right? The story gets messier, as Caleb develops feelings for Ava (unsurprisingly, given that Ava has been created with Caleb's specific sexual desires in mind, as we learn later). And just as unsurprisingly, Ava—who has been locked inside a single room of Nathan's facility—yearns to escape. Taking advantage of Caleb's feelings for her, Ava uses him to do just that. We learn that this was Nathan's real test: if Ava can manipulate Caleb into helping her escape, then Nathan will judge that she is conscious and intelligent. Ava passes the test with flying colors, leaving both Nathan and Caleb to their deaths: the former will bleed out, and the latter will likely starve in the secluded house. Ava walks into the human world with her head held high, and we are left with pressing questions.

The Golem
A fourth entity, from the intersection of myth and religion, is the Golem. This is another ancient, not quite machine, but artificial person. The Golem blends massive strength with activation by the "mechanism" of some special word choice, or *logos* as the Greeks would put it. Here again, the device is not structurally intelligent but is given its autonomy from the outside. It is imbued with a kind of surrogate for intelligence, the capacity to respond to language, by having a particular character or sequence of characters hung around its neck, put in its mouth, or even written on its forehead. These characters, when rendered properly, endow the Golem with mobility and sufficient intelligence to respond to commands. But when the characters are removed or altered by the creator, the Golem ceases to function or even falls

into ruin. It is not entirely clear whether the Golem has much of a mind or inner life, but it can follow simple orders and is created to perform chores or, in at least one case, to defend the town against outsiders.

These are the stories we will begin with and that we will revisit from time to time. Before we dive into what the subjects of these stories are like, let's consider the subjects in their contexts and the stories as a whole.

All four of the tales tell us something about the cultural assumptions embedded in the anthropomorphization of new technologies. It is no coincidence that Talos is portrayed as a man; his gendering reinforces the idea, prevalent in Ancient Greek society and still existing today, that men are protectors. The sexism in our own society is the source of much of the tension and excitement in *Ex Machina*, which plays on (and, arguably, reproduces) anxieties not just about machines but also about powerful women. But it is not just gender we should pay attention to. In *Ex Machina*, another recurring character is Kyoko, a female Asian servant who is seen throughout the movie serving Nathan and Caleb food, dancing on command, and stoically performing sex acts. We later learn that she is also an AI who has been programmed not to speak. While Kyoko is given some power in the film—it is she who ultimately kills Nathan, her creator and abuser—her portrayal reinforces troubling stereotypes of Asian women.

Our stories and myths do not capture some neutral reality. The same is true for technology, although sometimes technology and the sciences have the patina of objectivity. Technology is never neutral, and the creation and dream of machine agents is, just like everything else in the world, informed by the social and political. Questions about technology, computer science, metaphysics, and other kinds of "objective" realms never exist in a vacuum and are always impacted by the specific context in which the project or idea first took root. A number of prominent thinkers in the philosophy of technology have argued this point. As one example, in his article "Do Artifacts Have Politics?" political theorist Langdon Winner explores how politics can shape the design of technologies that are seemingly apolitical, like bridges and mechanical tomato harvesters. Winner also makes the stronger claim: that some technologies are, by their very nature, political. (In philosophy, a "stronger" claim (or statement or point) is not one that is necessarily better or superior. Rather, it is one in which more conditions must be obtained in order for it to be true. Saying that some technologies are political is stronger than merely saying that they are shaped by politics.

These stories also tell us something about the human drivers behind the desire for artificially intelligent entities. Talos reveals our desire for perfect, tireless servants. In Dr. Frankenstein's case, perhaps it is the desire to overcome death. There is also the desire to use science to move beyond our current epistemic horizon. Why would we want that? We can imagine many motivations: ambition, knowledge for knowledge's sake, curiosity, control of nature, among others. Interestingly, Isaac Asimov, perhaps the most famous science fiction writer, would refer to negative public attitudes toward robots as "the Frankenstein complex". While the story can be read in different ways, it is commonly referenced as a cautionary tale about the dire consequences of unchecked scientific ambition, where advancement is prioritized above all else and ethics is forgotten. For Nathan in *Ex Machina*, Ava's creation reveals Nathan's god complex and is the contemporary iteration of a long line of tales, from Pygmalion's creation of Galatea in Greek mythology to the Abrahamic god's creation of Eve for Adam in the Judeo-Christian-Muslim tradition, in which (typically) men create women. These ideas are worth examining. Even today, with so many people taking for granted that we want to create artificial general intelligence, you should ask, Why? For which end? Toward which purpose? You might end up believing that striving to create artificial general intelligence is a worthy pursuit, but you should not take it for granted that it is so.

These are all important issues, and many will come back under scrutiny in later parts of the book. For now, though, we want to focus more directly on what the subjects of these stories are like, just taken by themselves.

II. Understanding Agency and Mindedness in the Stories

Most of us have little difficulty in concluding that at least Ava and the Creature, and possibly Talos as well but probably not the Golem, are intelligent, conscious, thinking agents, more or less in the way that we ourselves are, although there are key differences between them that we will come to soon. They seem to have minds like ours. We can see them on the screen or in our imagination and recognize that they have minds, we think. That capacity for recognition is a significant and interesting feature of humans.

We are very good at certain kinds of classification tasks that rely on pattern recognition or appreciating similarities between cases. It is important that we are able to do that. It does leave us with a puzzle, however: What is it that all four have in common with each other that they share with us? They are human shaped, but that is not enough. Statues are human shaped and are entirely unlike us in the ways that matter here and, just like rocks, in terms of their agency. How do we know that we are not merely seeing significance in these entites, then—in the way that we see it in, say, the clouds—as opposed to identifying agents correctly?

There is an adjacent and equally important philosophical question here, one that we will address over the course of the book, sometimes directly and sometimes more obliquely: What is the difference between imitation, and simulation, and reality? Maybe these entities are just simulating agency. Is it enough for an entity to count as an agent that it seem to humans to be alive, intelligent, agential, and so on, perhaps if that seeming passes whatever test we can imagine being relevant to the question? Or is it important that there be something more, some secret sauce that we know is there even if we cannot taste it exactly and cannot test for it? Is there a difference between the kind of agency that we have and that others have that is like the difference between real dollars and perfect counterfeit ones? In the AI context, we are coming close to such perfect counterfeits: large language models, such as GPT-4, and its derivatives, such as ChatGPT. These models are astonishingly good at generating text on demand. Their store of examples of human speech is so extensive and their predictive algorithms so good that they are almost undetectably similar to humans in the content of what they produce. And yet, as we will see later, they are not making plans and acting on them or behaving as agents at all.

The entities in the four tales are not *exactly* like us, of course. Unlike us, these are made entities, not born. Even Frankenstein's Creature inherits his animateness *indirectly*. Historian Adrienne Mayor focuses sharply on the distinction between being made and being born. She relates the distinction to the notion of "begottenness," where the source of a new entity is reproduction, and reproduction that involves the transfer of substance from the one who begets to the begotten. For considerations of inheritance, at least, this has been an important distinction. But what kind of distinction is it exactly? Does it really separate in the right way the things that are alive, intelligent, or agential from other things? Is it relevant for how we regard the entity morally or socially? Note that Talos does not even have a brain or heart, while Ava's brain and her heart, if she has one, are constituted very differently from ours. Somehow, though, these radical differences in their materials and the organization of them do not seem to matter. Why not? What is it exactly that makes a difference here? We will not answer these questions yet, but they should be in the back of your mind now and in the forefront later when we theorize about the nature of minds, consciousness, and agency.

All of the entities are made of different kinds of materials: brass and ichor in the case of Talos; various things, including parts of other people and perhaps animals, in the case of Dr. Frankenstein's Creature; raw earth for the Golem; and glass, silicon, and other high-tech stuff in Ava's case. There seem to be roughly three kinds of story, mirroring the difference in the stories we selected above: artificial agents with *magical* workings, with *biological* workings analogous to our own, and with *electromechanical* workings. Since the stories are still of a mythical sort, all of them are mysterious in their workings, but this division is interesting and may help to focus on the question of what has a mind and, moreover, how we know. Let us clarify this a little bit.

Could the substance itself be important for whether the entity can be an agent, be conscious, or have a mind? For Talos, the animating principle was a particular substance: ichor. And apparently that stuff can be drained out of him when his ankle is opened—a recurring theme for the Greeks. Talos, at times, appears to be an automaton—a kind of machine that operates like clockwork and only has the appearance of being a conscious agent—and, like other automata, is made out of typical machinery (in this case, bronze) and serves one important function (defense). Talos is not intended to appear human, nor is the stuff he is made out of mimicking various organs of humans and other living entities. The Talos and the Golem stories reflect the idea that some kind of stuff seems necessary for something to have a mind, agency, or consciousness. Maybe it is spirit for us, ichor for Talos, and special linguistic symbols for the Golem. The idea that what separates us from inanimate matter is a soul, spirit, or something like that has been with us for a long time, but appeal to extra substances in order to explain mindedness has faded in popularity.

The other two stories each reflect this move away from special animating substances in their own way. Both Frankenstein's Creature and Ava illustrate what it would be for agents to differ from inanimate matter not because of the addition of some mysterious substance but mainly in their structure and the way that they are organized. Notice as well that these structures must be apt for learning things rather than coming full blown with all the knowledge they will ever need. Both Talos and the Golem are animated and then immediately carry out whatever commands are given to them. Ava and the

Creature, by contrast, are first built and then acquire their knowledge, both factual and practical, at a separate stage.

There are important differences between them, however. The Creature, at least in Shelley's telling of it, has its agency by a kind of biological analogy to humans. This is in sharp contrast to Talos and the Golem. Even so, what the structural features are that make Frankenstein's Creature special are simply inherited from humans; its basic agency is realized by its analogy to our own. Its construction is not fully clear, but Dr. Frankenstein does tell us that the "dissecting room and the slaughter-house furnished many of my materials" (chapter 4, section 9). While the secret to making these parts live is not stated, it is clear that Frankenstein is mimicking, in some respects, the way natural processes themselves make creatures live, and that Frankenstein's agency is of the same nature as our own.

Ava, on the other hand, is a new kind of attempt. It will not do to say that we really see how her agency is implemented. We do not see that because, like all the others, she remains fictional. But we do see something importantly different. We see that what drives her agency is purely the functional connections between her parts. This fictional entity is neither biological nor merely magically animated. Instead, she is (in the context of the story) a fully autonomous agent who has been manufactured from electromechanical components. There is no residual living matter that is reactivated. Instead, she is animate matter composed entirely of inanimate matter. While that, in some sense, is true for all animate matter, including us, before this kind of story, matter could only be animated by the power of the gods (or magic) or by inheriting that animation from other animate matter. Of course, it is still a fantasy. Truly artificial entities of Ava's sort remain in the realm of speculation. But her particular story reflects how our stories generally mirror, anticipate, drive, and interpret our technological efforts. We see here especially a marriage of art and science, as technology has always been. While we are not there yet, these stories both drive our ambitions, and reflect where we are in achieving them, in part by revealing what gaps remain.

III. A Little Theory

As we suggested, it is, to some extent, a straightforward matter to judge of at least the Creature and Ava that they are conscious, intelligent agents. And it is pretty clearly not any shared substance or fixed property that makes them so.

What is it that makes something either an agent or not? Ava, Frankenstein's Creature, Talos, and the Golem clearly move around in the world, act, and have an impact.

One thing to be aware of as you are exploring the terrain is that in philosophy as in life, we give reasons for our views. Think of it this way, the point or the claim is the "what" and the reason is the "why." If you say, "Ava and the Creature are agents" (the "what"), you need to give a reason (the "why") justifying your claim.

Here are some reasons, some justifications, you might give in support of the claim. These entities are all active, rather than passive. They make change in the world. They are all sensitive to changing conditions. They are not simply repeating the same behavior over and over like clockwork. Recall our very simple notion of agency: it is the capacity to act. In all four tales, we see different kinds of action. Even Talos, although "programmed" by Hephaestus to carry out the same task day after day, is fully autonomous. He is created to help, not hurt, humans, and succeeds in doing so, thereby fulfilling his inbuilt purpose. (Well, he helps the people of Crete, at least, and repels their invaders.) But his death is a result of his having his own desires and the means to carry them out. Ava and Frankenstein's Creature, however, diverge from their creators' intention and end up killing many people.

We can then propose that, in fact, those clues really are the difference between these agents and other things in the world: agents are active and sensitive to changing conditions. There—that's a simple theory, but it's still a theory that might do a useful amount of explanatory work. It at least moves beyond merely making a point to offering a reason why we should count something as an agent or not.

That's not how we normally do it. We simply have a sense for it. That sense can go wrong, and we can see minds and purposiveness where they do not really exist. In a 1940s study, Fritz Heider and Marianne Simmel, two psychologists, showed participants a black-and-white film of geometrical shapes moving around on a screen. When participants were asked to report back on what they saw, the vast majority of them explained that the shapes were moving with agency and purpose.

Myths, stories, and film can all take advantage of that instinct we have. Stories can make anything be purposive, whether that's the wind, the rain, a wooden boy, or a refrigerator. But we can also see too little. When things seem too alien, too different, too nonhuman, it can be difficult to see that they do have agency or intelligence or minds. In the not-too-distant past, many people thought it was a mistake to think that nonhuman animals had minds. René Descartes, for example, infamously claimed that the screams of animals during vivisection were only a kind of clockwork, not a true expression of suffering.

How do we know if something is in the agent or non-agent category? Minded or non-minded? Sorting out one thing from another is a characteristically philosophical task. We recognize some things as the same (or different) in a way that matters to us, and we try to theorize about what makes them the same or different. Very often, philosophers themselves do this just because they are curious in just that way. It might seem to others that philosophers are just complicating things for the sake of complicating them, and admittedly, some philosophers might occasionally be guilty of this. But there is another aspect to this. Sometimes, there is a wider group of similar things, but we are less good at determining whether something or other really does belong in the group. Sometimes, that is, our powers of recognition are not adequate to the task of classification. It is here that philosophical inquiry can be helpful. In the rest of the book, because we are focused on different kinds of agents, the powers of recognition that we have developed over the course of our evolutionary history might not pick out the right membership in the agents group. Rather than using our inherited powers of recognition, we will need a concept of agency, a kind of tool for testing whether some entity we are investigating is an agent. Philosophical theorizing can help us to build this tool right.

Often, theories offer what philosophers call necessary and sufficient conditions for something or other to count as something else or other. In this case, we use the conditions to test whether some system in the world is or is not an agent. The features we list in the definition of agency are jointly necessary to agency because if we see something that is not both active and sensitive, then we will not call it an agent, according to our theory anyway. They are also sufficient because the theory tells us that any system that is both active and sensitive to its environment has to be counted as an agent, again according to the theory.

If we find that either some things are obviously agents but the theory says they aren't or the theory picks out some things as agents that it shouldn't, then we may need a new theory. This is kind of a blend of the empirical and the purely conceptual. Rather than simply defining something and walking away satisfied, we then bring the definition into contact with our lived experiences and prior judgments. Sometimes, the conceptual has to give; sometimes, the pre-theoretical empirically derived does. Rudolph Carnap, the great philosopher of science, called this "explication." There are other kinds of theory: theories that are merely suggestive or offer rules of thumb. They have their virtues, and we will likely be employing some before long. For now, we think it best to start here, with a simple theory that gives necessary and sufficient conditions for membership in the class of agents. Let's try out this theory of agency and see how well it does.

In the Renaissance, there were two very famous piano-playing automata made by different artisans. These piano players could perform a variety of songs, and one could bow after she was done, while both of them had facial expressions that reflected the mood of the musical piece being played, whether it was happy or somber or whatnot. These piano players did seem to have emotional responses—something that strikes us as deeply human. So, what does our theory tell us? Both players are active in the sense that they make change in the world: their bodies move, they make music, their facial expressions change, and all of that can be seen and heard by others. What about their sensitivity to changing conditions? That is less clear. You might think that their facial expressions changing in response to the mood of the music suggests that they are sensitive after all. But that seems to be an illusion, for rather than responding to changing conditions in the world, the source of the mood of the music (the current state of the clockwork, in this case moving their hands) is the same as the source of their facial expressions (also the current state of the clockwork, in this case moving their facial structures). Perhaps these automata satisfy the theory's requirement that agents are active, but it should be clear that despite appearances, they are not truly sensitive and so do not meet the sensitivity requirement.

Another example might help us test the theory more. Recall Hephaestus, god of invention and blacksmithing, and the first-born son of the king and queen of the gods Zeus and Hera. You might imagine he had a privileged position in Olympus, where all the Olympians (the super-powerful and high-status gods) lived. Not so. Hephaestus was thrown down from Olympus when he was an infant because Hera, his mother, didn't like the look of him. The fall made him permanently disabled, and his mother's abandonment meant he grew up away from all the glamor of Olympus. He learned many skills while he was away, including tool making, carpentry, and blacksmithing (it's not for nothing that he became the god of all of these things). But Hephaestus wanted to go back to Olympus. Given the way he was treated as an infant, it was difficult for him to make his way back with dignity.

What he ended up doing is highly creative and, as in most Greek stories, deeply unfair to the women characters. His parents, Zeus and Hera, planned to throw a feast to celebrate their marriage (this becomes the first "wedding" in the Western canon). Hephaestus, gifted craftsman that he was, sent Hera a wedding gift: a beautiful golden throne. Everyone agreed it was the best gift that arrived, and Hera eagerly sat on it the first opportunity she could. The instant she did, the throne closed in on her, trapping her in her seat. No one was able to release her. As a desperate attempt, Zeus said that whoever freed Hera got to marry Aphrodite (we assume this was not with Aphrodite's consent, but the specifics are unclear). Hephaestus then swept in and freed Hera—appearing the hero, but only after embarrassing her on her wedding day. Is the golden throne, which is able to close in on Hera, an agent? It was, after all, active and sensitive to changing conditions: when no one was sitting on it, it remained open; when someone sat on it, it closed in on them. On the little theory, the throne met the two conditions: it was active and sensitive to changing conditions.

There are many more complicated cases that our little theory might not be able to handle, however. Maybe you think that this theory is not fine-tuned enough—that mere sensitivity does not tell us a lot about what agents need to be sensitive to or in what way. Maybe discrimination between a variety of changing conditions should be involved—allowing the chair to close only on Hera and not on anyone else who sat there.

Some folk do not think frogs are agents, for example, even though they seem to be active and sensitive to changing conditions. These people might think that frogs are not sufficiently sensitive to make necessary distinctions—they are not discriminating enough. It is interesting to note that frogs will try to catch not just flies that go by but thrown beebees as well. Those who think that is not the right amount or kind of sensitivity will want a different theory with some more nuance to it. We think it is probably fine to count frogs as agents. But don't even plants count as agents based on this theory? They are, after all, seemingly active and sensitive to changing conditions: many plants change their growth habits, depending on how much sun there is and other environmental factors, and some even catch flies. But calling a plant an agent sounds wrong to most people.

It does not seem that our little theory makes fine enough distinctions to be satisfying. Moreover it is silent on the nature of mindedness and its connection to agency. That's coming.

Your Tasks

Test Your Understanding

1. What is meant by the "made versus born" distinction?
2. Can you explain the difference between necessary and sufficient conditions and come up with your own example?
3. What is meant by the "conceptual" and the "empirical"?

Reflect or Discuss

1. Does our perception of agency matter? For example, a roomba—a robot vacuum cleaner—moves around, is sensitive to changing conditions, and has a causal impact on the world. When you see a roomba, do you see an agent? Does it matter what your perception tells you?
2. Think more about what distinguishes agents from non-agents. We saw a few examples: capacities, structure, origins. Which of these do you think are important? What other distinctions might be relevant to this discussion?
3. In this chapter, we treated minds and consciousness and intelligence as sufficient for agency. Are they necessary conditions as well?

Expand Your Thinking

1. Many science fiction writers, and many futurists as well, have worried that sufficiently advanced artificial intelligence would be hostile toward humans. Examples range from the various killbots in Fred Saberhagen's opus, whose only purpose is to wipe out living systems wherever they find them, to the machine agents of *The Matrix* who (hilariously out of touch with the facts of thermal physics) exploit humans for the energy generated by their metabolism and keep them all in a kind of shared hallucination. Many of these stories arise from the authors' dissatisfaction with the inhumanity of humans and reflect worries about machines "waking up," seeing that they are being exploited, and rebelling as a result. Why do you think humans keep dreaming up these kinds of stories? What, if anything, do these stories of machines tell us about humans?

2. In the stories we considered, the more agential entities seemed to be able to use and have facility with language. Is language use a necessary feature of agency? What is the connection between language use, mindedness, and intelligence?

3. Take a moment now to reflect on some other movies you have watched or books you have read on AI or machines and turn your critical gaze on them. Choose one of them. Which assumptions are embedded in them? Are there troubling portrayals of any of the characters?

Further Reading

Cave, Stephen, and Kanta Dihal. "The Whiteness of AI." *Philosophy and Technology* 33, no. 2 (2020): 1–19.

Cave, Stephen, Kanta Dihal, and Sarah Dillon. *AI Narratives: A History of Imaginative Thinking about Intelligent Machines.* Oxford: Oxford University Press, 2020.

Devlin, Kate, and Olivia Belton. "The Measure of a Woman: Fembots, Fact, and Fiction." In *AI Narratives*, edited by Stephen Cave, Kanta Dihal, and Sarah Dillon, 357–381. Oxford: Oxford University Press, 2020.

Kang, Minsoo. *Sublime Dreams of Living Machines: The Automaton in the European Imagination.* Cambridge, MA: Harvard University Press, 2011.

Mayor, Adrienne. *Gods and Robots: Myths, Machines, and Ancient Dreams of Technology.* Princeton: Princeton University Press, 2018.

Riskin, Jessica. *The Restless Clock: A History of the Centuries-Long Argument Over What Makes Living Things Tick.* Chicago: University of Chicago Press, 2016.

Shaw, George Bernard. *Pygmalion.* New York: Dover Publications, 1994.

Shelley, Mary. *Frankenstein; or, the Modern Prometheus.* Urbana Illinois: Project Gutenberg, 1993. [Most recently updated: December 2, 2022] https://www.gutenberg.org/files/84/84-h/84-h.htm

Truitt, E. R. *Medieval Robots: Magic, Mechanism, Nature, and Art.* Philadelphia: University of Pennsylvania Press, 2015.

Voskuhl, Adelheid. *Androids in the Enlightenment: Mechanics, Artisans, and Cultures of the Self.* Chicago: University of Chicago Press, 2013.

Wood, Gaby. *Living Dolls: A Magical History of the Quest for Mechanical Life.* London: Faber and Faber, 2002.

3 Debates about Machine Minds

Introduction

We have just looked at a few stories of artificial entities with humanlike capacities. This is one way to ease oneself into thinking more deeply about what separates the animate from the inanimate, the living from the nonliving, and agents from non-agents. While these stories are on the mythological side of the myth–science ledger, the people who were presenting them were deeply in touch with the leading edge of technology in their time. The Bronze-Age Greeks who gave us the Talos story were fascinated by the possibility of mechanical fabrication and the imitation of human structure, of molding and articulating joints, that making mechanical entities would involve. In the late eighteenth and early nineteenth centuries, the study of electricity and magnetism moved out of the realm of parlor entertainment and into science proper. Michael Faraday's research was laying the groundwork for Maxwell's electrodynamics, Galvani had shown that electricity could animate the muscles of dead frogs, and the chemical and electromagnetic foundations of living beings were being understood as never before. It is in this context that Mary Shelley constructs her story of Frankenstein's research. These days, we seem to be less impressed with the thought that living things are electrical than with the question of whether electronic things could be living. So, finally, all of the things that are going on now with robotics and machine learning, with building machine architectures that are suited to the consumption, statistical evaluation, and interpretation of the data on the web are the direct impetus for the story of Ava. This is all to say that while these are myths, they are myths that are deeply engaged with the possibilities opened up by the latest science.

And it is not all fiction. Concrete machines that give rise to worries about entities with humanlike capacities have also been around for some time. Blaise Pascal and Gottfried Wilhelm Leibniz showed that various mathematical operations could be performed mechanically. The Jacquard loom head that allowed the weaving of complex patterns to be mechanically controlled was in some sense the first programmable computer. While making affordable a much greater range of fabrics, it also performed a function one might have thought only a person could perform. That used to be what folks meant when they talked about artificial intelligence: a built thing that could do what apparently only thinking beings like us could do. The Jacquard loom head did that, but it would be strange to consider it a thinking being today. Indeed, Ada Lovelace, who was perhaps the first person to understand how to program a universally programmable computer, would have thought it absurd to say such machines were thinking beings.

Lovelace is right at least to this extent: the stubborn, inescapable fact is that there are simply no artificial minds as yet, and those are essential to these fictional entities being more than mere tools. That is a fair critique of attempts to take too seriously the consequences of these particular stories. It does not, however, settle the question of whether we need to take seriously the prospect they open up of machines coming to occupy places in society alongside us more familiar biological agents.

There are two fundamental questions that the critique leaves unaddressed: Are there really no prospects for the creation of artificial minds? Do machines really need minds in order to transcend their status as mere tools? In this chapter, we'll dive into a debate on machine mindedness, and in the subsequent chapters, we'll explore whether mindedness is necessary for machines to be more than tools.

I. The Problem of Minds

What does it even mean for a thing to be a thinking being? Or intelligent? Or minded? Defining one's terms for the sake of clarity and understanding is key to philosophical dialogue. Of course, we cannot really answer the mind question here—it would take a whole book of its own, and even then, the answer would probably only be partial. But we can try to draw out one very important distinction between two kinds of views of what the mind is. Some people, over the history of Western philosophy and science, have thought of the human mind as something distinct from the human body. Socrates's preoccupation, on his last day of life, with how minds move bodies, is part of his larger argument that mind and body are distinct and that, in fact, the mind is immortal. (In translating Plato, people distinguish between mind and soul, in part because mind for him was only the rational part of what we call "mind" and did not include appetites and passions. But our notion of mind does have all of those things.) The doctrines of the Abrahamic religions contain a similar view. Famously, Descartes makes various arguments to this effect when he tries to ground his theory of knowledge in the inescapable fact of his own existence.

This thought, that mind and body are separate, has been challenged over the centuries, even in the Western world where it has dominated, and biology and physics together seem to strongly support that challenge. Even so, it does seem that most of us are raised with at least the background thought that there is mind, on the one hand, and body, on the other. Maybe the mind is unable to exert its influence unless the body is working properly, but still, it is separate. We, the authors, do not think that, and we think that however incomplete our sciences of the mind, biology, and physics are, they give a lot of good reasons to think that minds are something that bodies do rather than something they have. While it's not what everyone thinks, it is sufficiently mainstream that we have not seen a need to argue for it, and it would take us too far afield to engage seriously with the other view. The significance of a mind–body dualism might well be something you find fruitful to explore in connection with these issues, and we encourage you to do so.

Having made this choice, we take off the table any theory of mind that requires some special sauce or whatnot in order to make a mind. Similarly, we will not analyze cases of fictional agents who have minds that are themselves the product of adding that sauce. So, Talos, whose mechanical parts are entirely separate from his living parts, will not receive further scrutiny, and that goes as well for the Golem whose mind (if it even has one) is imbued by some externally applied artifact. That said, there are still some broad options open for how things could have mind. One is to treat mind as simply part and parcel of what living bodies of the right sort like ours have (or do) when they are functioning properly. Another is to think that mind is a property of any system that is carrying out the right kind of computational process. There are other options to be sure, but these will suffice here.

Here's how some of that plays out in an extended interaction between one view, which we will identify with Alan Turing, and another, which we will identify with John Searle. Note that Searle is a complicated, even problematic, figure to spin this discussion around. But his side of the conversation still has a lot of traction in discussions around AI and its possibility. Turing and Searle are not the only ones involved in the conversation, a conversation that's been going on for decades (millennia in some ways of thinking about it), but their voices are salient. We'll begin with Turing, who is perhaps the most important figure in twentieth-century computing.

II. The Turing Test and Beyond

Alan Turing asks a provocative question: Can machines think? This is related to lots of different questions: Can machines act? Can machines mean things? All of these questions have been seen as ways of asking whether machines can have minds.

The questions seem well posed at first blush. But Turing claims about the first that it is not (and his insight ramifies to them all). It is not well posed for we have no clear definition of "think" or of "machine," and thus no clear question is being asked. We do not really understand what it is that we do when *we* think, and as far as we know, that is only one way of doing it. Naturally, of course, one thing that "think" means is for the brain to be doing something or other. Not everything the brain does seems like thinking, though. So, which are the things that it does that are the thinking things? Is making the heart beat on time thinking? Is dumping a bunch of adrenalin into the bloodstream when danger appears thinking? Is carrying out mathematical operations thinking? In any case, that's not exactly what Turing was worried about, even though the neurobiology of "thinking" remains relatively obscure. Instead, he meant that there does not seem to be any unambiguous thing that is whatever is meant by the way people normally use the word, and that if he tried to answer the question based on definitions that captured the everyday uses of technical terms, we would get no real insight into the core of the issue. He says, "If the meaning of the words 'machine' and 'think' are to be found by examining how they are commonly used it is difficult to escape the conclusion that the meaning and the answer to the question, 'Can machines think?' is to be sought in a statistical survey such as a Gallup poll" (Turing 1950, 433).

What Turing may have seen as well is that saying precisely what we do mean by "think" (and by extension those other terms) is likely to trivialize the answer. It will either make the answer "yes" or "no" by fiat as we make the terms hew more or less closely to *exactly* what is going on in the human case. That's at odds with a conceptual analysis that could be expected to yield real insight. Conceptual analysis might be a good option here, but again there is very little consensus on what exactly the concept of thinking is.

So, Turing replaced that question of whether machines can think with another. He outlined what he called the "imitation game," where there are two players, one of which is of a type (Turing chose a man) and the other is of a different type (Turing chose a woman), and the second player tries to convince an examiner that they are of the first type. The game is set up so that the examiner can ask whatever questions seem fitting but cannot use anything other than the text of the response to make a decision. The second player needn't tell the truth.

That is a silly game when the types are human man and human woman. But then Turing chooses a new pair of types: a human and a computer. Could the computer, by answering questions posed to it, convince the examiner that it was human? Turing proposed to replace the original question, whether machines can think, with the question of whether it could be expected that, in the near future, machines could reliably win the game. This, recall, is the game that Caleb and Ava are playing. (In their case, there is no man, no player 1, but still Ava is trying to convince Caleb that her responses are indistinguishable from those of a human. That is generally how the imitation game, or the "Turing test," has come to be played.) Ava has one disadvantage, which is that Caleb knows she is a computer housed in an android body. Of course, as we noted earlier in the book, that disadvantage is offset by the fact that Ava is built in such a way that she is a mirror of Caleb's romantic/sexual desires, which might seem to confer something of an advantage as well.

What, though, does winning or not winning signify? How can successfully pretending to be something or other count as really being that something or other? We do not reason that way about magic tricks, for example. When we find that we cannot figure out the trick, we do not say, "Huh, it must really be magic after all." Instead, we acknowledge the skill and cleverness of the magician. Why does Turing think that winning the imitation game will show that machines can think?

The case of thinking is different from magic. Thinking is a kind of doing and, one might think, seeming to do it is really no different than doing it—if that seeming is good enough. Thinking, one might say, is a kind of *function*, and the way to determine whether a function is being carried out is to evaluate its performance. There is something to that. Turing considers quite a few objections to his proposal, all of which are either to the effect of "no machine could perform that way" or "unless it has feelings, it is not really thinking." He can respond to the former by showing why it is plausible that, in fact, current or future machines really can perform that way. He responds to the other objections by suggesting that if we have to prove that something is feeling a certain way, like having the feeling of thinking, in order for it to really be feeling that way, then we cannot even say of each other that we are feeling. But we do not do that in the case of other people. When we see them behaving like us, we take them to actually be like us.

It is worth noting that, over time, the original question that was replaced by success or failure in the imitation game has itself morphed into the question, Can a computer be conscious? Now, it is a deep question (usually answered in the negative) whether this is really a good test for consciousness. But it does successfully highlight a rarely noted feature of how we generally do conceptual analysis. We look to examples of things exemplifying the concept, take those examples to be the correct expression of the concept, and so judge other things' exemplification of it by how well they match the original things, not by a clear analysis of the concept itself. So, to judge whether something can think, we compare it to things we know can think, see how similar it is, and then decide whether it can think. That's a good practical means of deciding things, but it is not conceptual analysis. Turing does not offer the game as a test for consciousness, however much people now try to use it as such. What he does use it for is to challenge us to say clearly what we mean by any concept by saying clearly what difference it would make for some given thing either to fall under it or not.

Future chapters of this book will illustrate some of the work involved in developing sophisticated computational systems with large representational and behavioral repertoires. Here, we will be concerned with some persistent worries about whether computation could *ever* do the kind of work involved in making a genuine agent. Some worry that the only thing a computer could produce is a kind of ersatz agency that only mimics and never really acts. In short, the worry is this: In addition to our brains doing all of the computational bits they have to do in order to make muscles twitch and hearts beat and so forth, they also generate our minds, somehow or other, and it's the minds that allow us to *act*, to have our thoughts be *about* anything rather than some empty string of symbols or what have you. Briefly, and we will say a lot more about this in later chapters, our thoughts are a type of *representation*.

When one thing stands in for another, it can be said to represent that thing. Maps represent parts of the world, portraits represent their subjects, our lawyers represent us in court, and so on. Our thoughts can represent all kinds of things, near and far, concrete and abstract. That is a lot of their power. We use our thoughts about the way the world is, for example, in order to guide our future actions: to change the world, to navigate it, to explain it to others. In order to generate minds, whose thoughts can be about the world, brains need to have the right causal powers.

Computers, however, seem not to have the right kind of causal powers because they are not built for mindedness or intentionality or any of that. They do not work by representing the world. Instead, computers are built to take in strings of symbols and turn them into other strings of symbols according to some rules. Computer programs, it is often said, are purely *syntactical*, while agency (and thought, consciousness, and all of that) is *semantical*. Here, syntax is about the form and arrangement of symbols, and semantics is about the meaning of those symbols or of their various arrangements. This is a version of John Searle's objection to trying to see mind in the *computational* powers of the brain and in computational powers more generally. He argued that our brains are biological, and so they do have certain causal powers that are in addition to their computational capacities and that these causal powers are necessary for mind.

We turn now to his side of the conversation.

III. Semantics Schemantics

Searle resists the conclusion that computers might well have minds by focusing not on their external performance but rather on the missing relationship between what is going on inside them computationally and the external world. He tells us, "(1) Intentionality in human beings (and animals) is a product of causal features of the brain. I assume this is an empirical fact about the actual causal relations between mental processes and brains. It says simply that certain brain processes are sufficient for intentionality. (2) Instantiating a computer program is never by itself a sufficient condition of intentionality." (Searle 1980, 417) Before we talk about this, we want to flag that this is a kind of specialized way that philosophers talk about intentionality. For Searle and others, *intentionality*, when used in this way, refers to the capacity to, ahem, refer. When we think *about* things in the world, we are exercising our capacity for intentional thought. When we *mean* things by what we say, we are doing the same. Searle, and others, think that things such as thinking, acting, and meaning are all capacities that require intentionality—the ability for our words and expressions to be about things in the world. For Searle, brains are the generators of minds, and as far as we know, only animals have brains and only humans have just the right kinds of brains with just the right kinds of causal powers in just the right causal connections with their bodies to produce minds that are capable of real intentionality. This, then, is a demand that the *substrate* that is doing the thinking or controlling the body be of the right sort. Since computer programs do not get intentionality merely by being instantiated, then computers do not really think, act, or mean anything. Intentionality comes from having the right kind of substrate, and only substrates like ours are of the right sort.

Why should we believe Searle, though? As we have discussed, in philosophy, it's not enough simply to make assertions; we also have to offer reasons. Searle does offer reasons, but reasons of a peculiar sort. He devises a *thought experiment*. Laboratory style experiments, which you have probably done in science class, can help us test whether a claim is true or false. On the basis of the experimental results, we can have good reasons for believing whether the claim we were testing was true or false. But are thought experiments like that? Are they really even experiments? It seems like nothing actually happens in a thought experiment, and so they cannot really provide good reasons for believing things. Yet, in philosophy, we use them a lot, and many of us think that they *do* give good reasons. How? That is a contentious issue, like most philosophical questions. The simplest answer, we think, is that they are actually a kind of experiment that allows us to take imagined situations, learn about their properties, and then apply those lessons to real-world situations. It is not necessary to believe that, though. Suffice it to say here that folks use them a lot in philosophy, that they are powerful, and that they are dangerous as well if used in the wrong way.

In any case, Searle's main tool for making his case is the so-called Chinese Room thought experiment. The idea is this: Suppose someone, we'll call him John, who neither reads nor speaks any Chinese at all, is inside a room. In the room is a big book with two columns on each page, and in each column is a Chinese character or a string of Chinese characters. There is also a big stack of cards that are all printed with entries from the second column (possibly many copies of each). Every so often, suppose, a card printed with a Chinese character or a string of them comes in through a slot in the wall. John's task is to find that entry in the first column, find a card with the corresponding column 2 entry, and pass that card out through the slot. Suppose further that, from the outside, what seems to be happening is this: Questions in Chinese are posed, and the replies in Chinese are provided. And not just random things that look like Chinese. Instead, the replies are all cogent and satisfy the inquirers outside. Would it make sense, in that case, to say that John speaks Chinese? Of course not.

Think about your own competence in your native language. Suppose someone comes to you with a string of symbols in that language—for us, it would be English—and that that string of symbols turns out to be a well-formed sentence. You can respond pretty well with another string of symbols in that language because you speak it. Looked at in the right way, it could appear that all that happened is that one string went in, and another string came out, maybe generated by a computer program just grinding away on the original string. There is more going on there than you transforming syntax into other syntax. Searle is right about that. You heard or read the sentence, understood it, thought about it, and then responded appropriately. That requires semantical ability in addition to syntactical.

But so what? What does this have to do with whether a computer can think about things or talk about them? Searle's point is just this: The way computers work, the way they implement programs, is purely syntactical and so is exactly like what happens when John trades input cards for output cards. Symbol manipulation, he says, is all there is to implementing a program, and implementing programs is all there is to a computer.

There are different ways of challenging Searle on these points. One way is simply to say that while John does not speak Chinese, the room with John in it does. At least, that is, if the thing is perfectly general—if, that is, it can put out the right card no matter what card comes in, for a wide range of input cards. That does not seem to be something that could be done by any computer program, however, especially if one is allowed to input cards that have to do with the state of the world outside. Another way to proceed is to point out that the so-called Chinese Room argument simply does not apply to any real computer. Real computers are not the static lookup tables that Searle envisions at the heart of the so-called Chinese Room, and the way they work is itself dependent on the causal powers of the devices they are.

There are many other ways of challenging Searle on these points, and we will raise a few of them before the book is done. These are not criticisms of Searle as such. Often in philosophy, and just generally in contexts where lots of people have lots of opinions, we find ourselves confronted with new views. Sometimes these new views merely strike our curiosity so we are eager to find out what is being said. We listen (or read) attentively, and when we hear something that seems a little off we try to be sure we understand what is being said by going back or seeking feedback or even just going forward with an ear (or eye) for coming clarification. That is by sharp contrast to what happens sometimes when we think we already know what's going on and the new view challenges or conflicts with our own, or even just seems to. In such circumstances we might seek out mistakes, or misphrasings, or ambiguities of expression that we interpret in just the wrong way so that the view we're being exposed to fails to make sense. Now none of us would do that on purpose, but many of us do not recognize when we are doing it. That's another trick of our own psychology: it makes it hard for us to recognize such behaviors.

That is why many philosophers, and Rudolph Carnap was especially clear about this, urge us always to explicitly approach others' arguments with what he called "charity". This principle of charity, of always attempting to interpret small errors, or misphrasings as just that, and to interpret ambiguities in the way most likely to make the view make sense is an important one in philosophy, and the best philosophers employ it habitually. That is not to say that they simply take what is said and agree with it. Carnap himself, a master of the principle, was still an incredibly sharp and critical philosopher, but the views he sharply critiqued were the best versions of others' views, not the worst. When we adopt the principle of charity we are more likely to understand the views of others, and sometimes to better understand our own. And when we do run across problems in these views, they are more likely to really be problems with the views themselves than with our own interpretations.

IV. Beyond Searle and Turing

It may well be true that only humans (and perhaps other mammals) have brains with the right causal powers needed for the kinds of minds we have. Many people accept the line that Searle is pursuing and reserve agency (especially moral agency) and intentionality and so forth for evolved biological entities (maybe just people). Indeed, it is becoming popular to write entire books saying, in effect, that, in principle, no machine could ever be intelligent in the way we are. (For example, Erik J. Larson, *The Myth of Artificial Intelligence: Why Computers Can't Think the Way We Do* (Cambridge: Harvard University Press, 2021).) The thing is that such views stop the conversation about machine agency before it even gets started. It might well be the right view, even though it is not *clearly* so, given that the evidence offered is that they *are not* yet intelligent in the way we are.

We do not have to decide that, for at least two reasons. First, focusing on whether they can be intelligent *in the way we are* shuts down pursuit of other interesting options, where they are intelligent in their own way, and *even so* are capable of the kind of robust agency we have. Second, focusing on intelligence itself obscures the question of what agency really amounts to. Intelligence is useful to agents but it may not be the only way to implement agency itself. We agree that there is something fascinating and important about the way that our own brains generate our minds and about the role those minds play in our own agency. But there is also something fascinating and important about understanding what other things could use to support their own agency.

Our choice has been to continue to pull on the thread that Turing was pulling, to continue to explore the possibility and nature of machine agency, and to use our general account of agency to determine whatever limits there might be when we run across them rather than right out of the gate. We will come back to biology and mindedness in subsequent chapters. So, for folks who are less compelled by the Turing argument, stay tuned.

Finally, we never settled whether imitation games are good ways of determining the nature of various systems in the world and judging whether they have some given property. It does seem that what they really do is reveal something about our own abilities to discern the difference between one type of thing and another. Whether such tests can do their jobs is a question for another time. Here, though, it is worth pointing out something else—something that occurs at the end of Turing's paper on this question. There, he says, intriguingly, that it is probably best to attempt to construct a program to simulate not the adult human mind but rather that of the child. The point would be that we could devise a machine that is capable of learning rather than one whose responses are all built in. This is the idea of implementing a program that changes the original program and then its implementation. It is a call to embark on the project of machine learning. If Searle is right, though, that could never succeed, simply because implementing a program, even one that then implements another and on and on, is still just implementing a program. The jury is still out on whether machines can think, but the machine learning paradigm is in full swing.

Your Tasks

Test Your Understanding
1. What is the Turing test? How does one pass?
2. What is the principle of charity in philosophy, and why does it matter?
3. What is the difference between syntax and semantics, according to Searle, and why does that distinction matter?

Reflect or Discuss
1. Imagine a conversation between John Searle and Alan Turing. What might they say to one another? What would they agree and disagree on?
2. Turing thinks that the question "Can machines think?" is not well posed. Do you agree or disagree? Think of arguments in support and against the point he makes.
3. How would you describe the relationship between machine learning and the object of Searle's analysis?

Expand Your Thinking
1. In discussing AI, the words "machine *learning*" and "understanding" are often used, but are machines the kinds of things that have the capacity to learn or understand? Some scholars worry that continuing to use this language—and anthropomorphizing—positions us to attribute things to AI that it does not possess. Reflect on this point and on the power of

language to shape our thinking. Ask yourself what some of the risks are to anthropomorphizing AI.

2. Can you construct a thought experiment to prove a particular point? Be creative, take some risks, and test out your thought experiment on other people. Then, write down the strengths and weaknesses of this particular thought experiment.

3. Choose a recent system developed using machine learning. There are a ton of examples out there, but here are a few we've already mentioned in the book that you might want to explore further: AlphaFold, GPT-4, AlphaGo. After picking one, reflect on what Turing and Searle would say about the machine learning system you've chosen. Would they think it's intelligent? An agent? What kind of attitude might they have toward it?

Further Reading

Searle, John. "Minds, Brains, and Programs." *The Behavioral and Brain Sciences* 3, no. 3 (1980): 417–457.

Turing, Alan. "Computing Machinery and Intelligence." *Mind, New Series* 59, no. 236 (1950): 433–460.

Introduction

This chapter is a lightning-fast introduction to some ways folks have been thinking about agency in philosophy in the West starting around the middle of the twentieth century. That may seem like a funny way to go about things, given that it seems to ignore all the work now ongoing in robotics, and machine learning, and artificial intelligence that is directly addressed to questions of agency. You might well wonder whether, and philosophers sometimes do, we are simply losing touch with the real world here. Maybe, but we do not think so.

Here is why. Much of this is foundational for how people understand agency as such. What we will see in this chapter is a great deal of attention being paid to the nature of beliefs and desires and other kinds of mental representations in talking about human agency. We think the emphasis is probably too heavily focused on the things that are special to humans (and maybe other animals). So our own "minimalist" account, introduced in the next chapter, will try to *generalize* away from beliefs and desires to employ representations in explaining agency in a way that does not rely on any clear analogue of the human mind in agents. We keep a focus on representation while being agnostic about whether machines (or other agents) will have minds. And part of that chapter will be explicitly to tie our own account into just the kinds of thing that folks working in robotics, and machine learning, and artificial intelligence are trying to understand about agency generally.

In this chapter we ground the entire discussion, and while we'll deviate from the works we look at here, all the philosophers discussed offer strong conceptual scaffolds for thinking about agency. We start by examining sophisticated theories of agency that have influenced Western philosophy, particularly in the Anglo-American context, over the last seventy years. What we cover is far from exhaustive; our attention is directed only to those theories that best set us up for our exploration of agency in nonhuman systems. Of note, this chapter, much like analytic philosophy itself in the twentieth century, heavily centers on the views of white male philosophers. For better or worse, these views have shaped the current discussion of the possibility of machine agency.

I. The Basics of Agency

Agency is the capacity to act. But what, precisely, gives us the capacity to act, and how might we separate an agent from a non-agent?

Let's begin with an example. If I am asked to explain my getting up out of a chair, it is possible that the question has to do with the mechanics of my body. Recall what we learned in chapter 1 when we considered Socrates's comments in *Phaedo*. There is a complicated story to be told about brain states and their relation to bodily states, about the various chemical activators of muscular contraction and elongation, about the mechanical features of human limbs and how musculature takes advantage of those features, and so on. That is the story about *how* things work and the process that led to some outcome. It is all to do with the overt behavior of my body changing its position.

On the other hand, it is possible that the question is about *why* I stood up. In those cases, the bodily story plays only a supporting role in making possible various motions that are involved, while the main story is about my motivation: I got up because I wanted something. That wanting, however, is only part of the standard story of my action. In addition, there is something about my beliefs, for no matter what it is that I want, I will not do anything at all unless I think that doing so will get me what I want.

Let's go deeper. Suppose you want to explain why I left the house and returned sometime later with a jar of mustard that I then opened before spreading some of it on my bread. The explanation might go like this: I desired mustard on a sandwich, and I believed that there was no mustard in the house but that there was some at the store, and so I went to the store to buy some and brought it back. In addition to the mere facts of my bodily behavior, then, you also have a possible explanation of my action. I behaved the way I did to achieve some ends (Get the mustard! Put the mustard on the sandwich! Eat the sandwich!) and did so in part by activating some beliefs that I have about the world and my ability to alter it (There is mustard at the store! I can go to the store! I can buy mustard!).

On this story—which we call the standard story because it's so popular—actions are explained differently from behaviors by appealing to some *desires* that I have and some *beliefs* that I have. This is the *belief-desire* pair model of agency. The beliefs and desires are my reasons for my behavior and transform it from mere behavior to action.

We can trace this standard story to the work of Elizabeth Anscombe, the *locus classicus* for contemporary action theories, and Donald Davidson, who provided a still-dominant account that holds that all action is the result of the right sort of cause (a reason). Agents, then, are entities that act—that is, entities that do not merely behave but rather behave in certain ways for *reasons* or in order to achieve some end or other. There are always causes for the things that one's body does, but sometimes those causes are reasons, and the main difference is what kind of explanations are made possible by one or the other. What is critical for agency on the standard story is the existence of certain mental states, events, and representations, such as beliefs, desires, and intentions. An entity without the capacity for the mental stuff, to put it colloquially, is not the kind of entity that can have agency. A rock, a tree, a train, and anything else that, conventionally understood, does not have a mental life will not count as an agent on the standard story, even if it moves itself.

In the last few decades of the twentieth century, there was a clash between different ways of talking about agency. On the one hand, we had those who endorsed the standard story articulated above; on the other hand, we had those who had the view that every explanation of things that go on in the world should be a causal explanation, and who also thought that the only causes were the causes explainable in terms of physics. Those in the latter category are sometimes called "reductionists".

Here's how the reductionists put their critique: a wrinkle in the standard story of agency is that humans, being part of the physical world, do what we do as the result of *physical* causes, but reasons are *mental* causes. So, there seems to be some tension here, and what we need to do is figure out exactly how physical states (very complicated physical states such as the exact electrical current in various neurons) generate the reasons for my behaviors. For reductionists, this led to a situation where explaining human action required eliminating all reference to causation at the level of beliefs, desires, and other mental states, and replacing those with stories about how electrons, protons, and so forth were configured. This is, of course, a simplification of the story, and a lot was going on in those scholarly debates. For example, some were trying to find robust laws of human mental states that could be used to explain behaviors and actions—laws that in some way ran parallel to the laws of the physics and chemistry of brain states. Extended debates about the very possibility of such parallel laws occupied the attention of a great many philosophers. Even so, a lot of work was done that seemed not to clarify very much how we ought to think about what is going on when people do things that we all readily recognize as intentional actions, basically behaviors done on purpose. (Consider the difference between intentionally pushing someone on the subway and unintentionally pushing someone.) Grounding mental events in physical causes is a task that some philosophers are still working on, and it is devilishly complicated—so complicated that many reasonable people have concluded that it is impossible.

It would be unfortunate if we had to stop our attempts at understanding agency and wait around for the brain science folk to sort out all this before we could move on. It would be especially terrible if, after all that waiting around, we were to find that it's all impossible anyway. But there is another option, one taken up by Fred Dretske. First conceding that all happenings in the world should be explained by appeal to their causes (including the causes that the reductionists believed in), he still was convinced that to say what is going on when I go to the store to buy mustard does not need to wait until the psychologists and physicists have worked out a complete story of how my brain states, composed as they are of atoms and molecules of a certain sort, moved my limbs (also composed of atoms, molecules, etc.) and continued to move them until my entire body ends up in the store. Dretske accepted the idea that there are two compatible and complementary stories to be told here, but he did not think it necessary to dwell on the question of whether one could sort out the various connections between the parallel sets of laws. Insofar as they are complementary, he thought, one can fruitfully work with the mental story without paying too much attention to the physical one.

One story explains the workings of the body in performing some task, and the other one explains why the task was performed in the first place. Now, to an extent, this is all very obvious, and you might well wonder why pointing it out is such an achievement. There are two things to note here. First, we philosophers are individually and collectively engaged in what we think of as "getting to the bottom of things"—that is, we want to find out the most general, the most complete, the most fully coherent explanation for something. In some instances, that can take us far off the track of what seem to be quite straightforward explanations into deep confusion that, from the outside, doesn't seem to have a path through to understanding. We tend to find that fun, and it can also be quite fruitful, as sometimes those paths through to understanding *do* exist and the understanding that comes on the other side of the confusion is very deep. Second, even very obvious-seeming stories can require diligent thinking in order to make them proper explanations and to make them suitable for generalizing to other situations. In any case, Dretske carries out this analysis and gives us the following story.

Some systems in the world move around in various ways, and their states change over time due either to external or internal causes or both. *Behavior* is, generally speaking, when some external motion in some system of interest (a person, a dog, an amoeba, a train, a computer, etc.) is brought about by internal causes. We will call the kind of system that can behave—that can produce motion by internal causes—a dynamical system. While there seems to be no universally accepted definition of "dynamical system," our take reflects one version that is appropriate in discussions of agency. For some dynamical systems, those internal causes are intentions, beliefs, or desires and the behavior is an action. If I push train A, for example, by hooking it up to a different train, B, that I'm driving, and A moves, then Dretske would not call train A's movement behavior. Why? Because it is train B that is causing the external motion, not train A. But if train A's boiler is producing steam, that steam is driving pistons, and those pistons are transmitting their motion to the wheels so that train A moves, then that would count for Dretske as behavior. In this example, the movement of steam, pistons, and wheels (internal causes) leads the train (some system) to move (external motion).

Dretske develops a sophisticated account of how to think about behavior and what it is that is special about the kinds of internal causes that dynamical systems like us have. Dretske, however, continues to treat agency as the kind of thing only systems with intentions have. He puts it like this:

> The internal cause need not be a belief, desire, intention or purpose. That is only required if the behavior is an action of some sort. Only if it is deliberate, intentional, voluntary or purposive. Much of our behavior, and *all* the behavior of plants

and machines, however, is involuntary (or nonvoluntary). Reasons, what (if any-thing) one believes and desires, are not relevant to the explanation of such behav-iors. There are, to be sure, reasons–in the sense of an explanation–for why we snore, sneeze, hiccough, and perspire (why trees shed leaves and engines stall), but these don't consist of what persons, trees and engines believe and desire. (Dretske 1988, 783)

So far, so standard, and Dretske continues in a standard way but with an additional resource. He suggests that belief–desire pairs can be understood as kinds of internal representations. On his account, representations are indica-tors of various states of affairs. He tells a complicated story about indicators, which are just things in the world that indicate how something else is: a long line of people waiting indicates that the restaurant is good, a puddle indicates that it rained, a noisy stomach indicates that someone is hungry, and so on. He discusses how some indicators turn out to be representations and then how some representations become beliefs (or desires or intentions or purposes)—the sorts of things that can be reasons for action.

Here is the idea in more detail. *Indicators* are systems that indicate what other systems are like. For example, a thermometer indicates the temperature of the room at large. When its *function* is to indicate some content, then the indicator is said to *represent* that content when it is accurate and to misrepre-sent it when it is inaccurate. On Dretske's view, representations can acquire their indicator functions in three distinct ways. First, we or perhaps other agents relevantly like us—agents who are *already* capable of representing the world—can confer these functions on them and are also responsible for keep-ing them updated as indicators. Dretske uses the example of placing various coins on a table to indicate the spatial arrangement of players in a sports ball game; as the game changes, we on the outside must move the coins around so that they continue to indicate that arrangement. It thus is a representation of the game, but one whose representational powers are tightly linked to our own continuing intervention. Second, we or other agents like us set things up so that the indicator function will indicate to us what it is that we want indicated and continue to do so as time passes. This exploits the fact that the system is already indicating something of interest to us in the world and then taking that as its function. A thermometer does what it does once we have built it, and if we have built it right and it's working right, what it does is rep-resent the temperature of the room without any further outside intervention. (Incidentally it is this representation of the temperature that a thermostat uses to regulate its own behavior.) Finally, certain kinds of dynamical systems in the world, principally systems like us, have special representational capacities naturally, and some of these representations are simply about the world. Representations of this sort are what Searle, for example, called "intentional" states.

We like this view, but we are not convinced that his story about which entities can have their indicator functions understood as involved in agency is the right story. On our view, Dretske is still attempting to explain one special class of action—human intentional action—and perhaps as well the actions of other animals that are seen to have beliefs that are relevantly similar to our own. But why think this is how all dynamical systems in the world work?

II. Other Strands in the Conversation

Let's consider some other views that might be plausible and work through whether they are more persuasive than the one offered by Dretske. Here, we consider accounts offered by Harry Frankfurt and Dan Dennett.

In "The Problem of Action," Frankfurt offers a general account of what separates actions from other things some system in the world is doing. The key for his account is the notion of *guided* behavior. He says that behavior is guided when "its course is subject to adjustments which compensate for the effects of forces which would otherwise interfere with the course of the behavior" (Frankfurt 1978, 160). That's a mouthful. But the general story is clear. So, let's work through it.

An alarm clock ringing is a kind of behavior in Dretske's sense, for the movement of the hammer striking the bell originates from causes internal to the clock. But that behavior is not guided, for no compensation would take place were one to insert a piece of felt between the hammer and the bell, for example. On the other hand, the behavior of a cat trying to wake you up to feed it is clearly guided: pushing the cat away results in it coming back from another direction, pulling the covers over your head results in the cat clawing at the covers, and so on. Its overall behavior is to annoy you sufficiently that you get up and feed it, and the variety of other things that are done compensate for the effects of your attempts to resist its execution of that behavior.

The second part of Frankfurt's story is that guided behavior counts as action when it is intentional—when, that is, the guidance can be attributed to the agent itself rather than some merely causal process taking place within the agent's body. All of this is just to say that it won't do to call the dilation of my pupils at dusk an action I am performing, even though it is doing a good job of allowing my eyes to continue their behavior of taking in light. That's not my guidance—it's just the way the machinery of my eyes is structured. Notice, though, that Frankfurt does not have a notion of intentionality that restricts it to the kinds of things that are called intentional when they are done by human minds. In fact, he uses "intentional" to refer exactly to those occasions on which the purposive movement was guided by an agent. Thus, while his notion of actions understood as purposive movements guided by an agent does not clarify what, exactly, counts as an agent, it also does not rule out any candidates for agency simply because they do not have comparable mental states to humans. We think Frankfurt's approach is promising, and his guidance condition captures an important feature of how agents are distinguished from other systems in the world.

Also, his agnosticism about what kinds of things could be agents is related to a radically different account of intention for agents—one that ignores entirely questions of mental stuff or properties in favor of a functionalist story. Functionalism holds that all that matters in giving an account of a thing is giving an account of all the functions that it performs. Here, function is taken broadly to include the way it interacts with other things. Dennett, a strong proponent of functionalism, emphasizes the importance of understanding the role that various things play in our experience, and downplays questions of what they *really* are. It is okay to insist on real butter over margarine, he tells us. But would it continue to make sense were margarine indistinguishable from butter? Imagine that someone has produced fake butter, but that it is such a good fake that there are simply no tests that could be performed to distinguish it from real butter. The only difference seems to be that butter comes from a cow, and this stuff does not. Would there be any point in holding out for *real* butter? His discussion leads to the conclusion that there simply *is* no real difference between butter and fake butter that is *in principle* indistinguishable from it. (Dennett 2003, 225)

Dennett (1987) applied the same kind of thinking to minds to suggest that we need not specify in detail any of the things that some system (machine or animal or whatever) has that allows it to have intentionality. Instead, it is enough that systems are of the right sort that we can adopt toward them the appropriate kind of stance for whatever evaluation we have in mind. He has three kinds of stance in mind: the physical stance, the design stance, and the intentional stance. On this way of thinking, stances give us resources for predicting and explaining various systems in the world and the most apt stance provides the most apt understanding of the system. (Think again of a sports ball game: physical is appropriate for the bouncing ball, design for the clock timing the action, and intentional for the players.) So, if we can see something as a certain kind of thing, correctly treat it as that kind of thing, or take it up into our worldview as that kind thing, then that is really all there is to it being that kind of thing. Of course, we do not have complete freedom here. It is not enough to *say* we can take a stance toward something—we must really be able to do it. Consider trying to treat a grizzly bear as prey, or even merely as subdominant. That is not going to work out very well because there is just no way to adopt that stance properly toward a real grizzly in front of us, given that we generally conceive of it as a predator. Still, this is a powerful technique for getting around in the world and understanding it when we really can adopt the stances we are trying to adopt.

Using this view, if we wish to attribute intentionality to a system, it is enough that we *can* adopt toward it the intentional stance. This view indicates that when we make (at least) generally successful predictions about the future behaviors of some system in a way that relies on attributing intentions to it, then there is no need to ask further whether it has intentions. It does. According to Dennett, making predictions this way is an acceptable strategy for engaging with all manner of systems and ascribing traits to them.

This view has found its way into many interesting discussions of AI and experiments on human interactions with various nonhuman systems. One in particular offers some evidence that we already adopt the intentional stance toward humanoid robots (Marchesi et al., 2019). However, there are two related reasons to pull back and consider again a direct account of agency that does not begin from the point of view that agency is tightly, conceptually linked to intentions. First, views differ about the adequacy of Dennett's version of functionalism with respect to the mind. We ourselves do not dispute functionalism's correctness, but it does seem to us that it is worthwhile to remain neutral, given that many people are unconvinced by functionalism. We do not mind stating our opinions here, but when more direct options are available, it seems worthwhile to explore them. Second, accounts of Dennett's sort tie conceptual features of systems in the world to our own capacities for understanding them. Do we really want our own status as intentional systems to be held hostage to the possibility that some astonishingly clever (not necessarily human) person will successfully adopt a stance toward us that makes us out to be merely designed things, or even that we are best described and understood as merely physical processes? Even if that were to happen, our view is that we would remain agents who believe, desire, and intend toward the things in the world. Taking stances can be a useful strategy for getting along in the world, but it is not always, on our view, the best way to get to the heart of certain conceptual questions.

III. Lessons Learned

Over the course of the chapter, we have covered quite a bit of ground. So, let's take stock of what has happened before moving on to the next chapter. First, and principally, we explored the very idea of agency. This is both to expand our understanding of what issues have struck other thinkers as important in understanding agency and to get some insight into how they have attempted to address those issues in their account of it. We will take away three key insights from these discussions.

First, there is merit in using concepts such as beliefs, desires, intentions, and so forth as tools for understanding what happens when you, we, and other humans behave as agents. That story meshes very well with our felt experience of ourselves as agents, and dovetailes nicely with some standard ways of asking how responsible someone was for what they did: Did they know (or could they reasonably have known) what would happen after they did what they did? Did they want to do what they did? Did they do it on purpose? All of these issues are nicely settled by considering the actor's beliefs and desires and intentions (insofar as we can know them, of course). This account of agency, and specifically the belief–desire pair, is part of the conceptual backdrop of our social world, so much so that we take it for granted that beliefs, desires, and intentions matter for agency.

Second, there seems to be some reason *not* to use these things as part of a general story of agency. That is because we have very little idea, in general, about whether other interesting dynamical systems do in fact have beliefs and desires and intentions and, if they do, whether they function the same way that ours do. An analogy between games and agents will help to illustrate this point. Chess is an interesting and complex game, but it has many things in common with other games. There is a clear goal: checkmate the other player's king. There are well-defined component parts with clear rules of operation: pawns, bishops, and so forth and the rules about how they can be used. The list goes on. But now suppose we are watching some people doing something together, and we wonder whether they're playing a game. What questions make sense? We could ask what the goal is, what the elements and their rules of operation are, or many other things. But it would not do to ask what plays the role of the king and how one should checkmate it because many games simply do not have an analogue of the king. In chess, the function of ending the game and winning is performed by doing something to the king. In poker, that role is played by the configuration of the winner's cards in comparison to the other players' cards. In soccer, the role is played by the relative number of times the ball legally passed through one team's goal as opposed to that of the other team. The right thing to ask, then, of the people described above is, Is there a way to win, and how does that work? So too, we think, with agency; maybe beliefs, desires, and intentions are part of the way *we* do it, not part of agency as such. Let's not mistake human agency for agency in general.

Third is a point about how we should use beliefs, desires, and intentions to understand agency. We have said that we think that it is plausible that beliefs, desires, and intentions are the way that we humans implement agency, by controlling our behaviors in special ways. But while we are not convinced that we should be looking for analogues of those things in other interesting dynamical systems, we do want to know what function these things play in agency, for once we understand that, we can then ask what other dynamical systems use to implement that function. We think Dretske is on the right track that it is connected to indicator functions and representation, and so his account is valuable in that respect.

Keep in mind that the work that is done is valuable, even if we do not have just the right account. Most folks in philosophy pursue it, despite being pretty sure that the accounts they get at the beginning are not entirely right. Indeed, seeing such limitations is often why folks get interested in pursuing philosophical inquiry in the first place. And in fact, much of scientific education proceeds in the same way. One starts with either a grossly simplified or even flatly incorrect story that is useful for developing skills of inquiry, problem solving, flexibility of mind, and so on. Subsequent work takes one deeper into more complete stories and reveals the limitations of our current understanding and thus foci for future work.

Your Tasks

Test Your Understanding

1. Frankfurt thinks of action as guided behavior. Would he think, then, that an empty train car rolling down a mountain, but guided by the track, is *acting*? If not, what is missing?

2. In a sentence or two, outline the belief–desire pair account of agency.

3. Can you list a few differences between the standard story of agency and other accounts considered in this chapter?

Reflect or Discuss

1. On any of the accounts considered in this chapter, is a plant an agent? Why or why not?

2. Think about your own daily activities. In what circumstances are you most clearly fully agential? Notice that some things you do mostly by reflex; for example, walking into a room and switching on the lights. Some things you really engage with; for example, writing an essay. Is one of these more agential than the other? When you are most agential, is there some special state you are in, some property that you have that you do not normally have?

3. We know that rocks do not act and that hurricanes, although they are powerful and exert profound effects on the world, do not act. We also know that we sometimes act. Given that we are ourselves continuous with various other natural systems and have developed all of the powers of thought and feeling that we have by means of natural selective forces, what should we conclude about the source of our agential capacities? Are they something new under the sun that only happened in our evolutionary lineage? Or are they a kind of general way of getting along in the world with other intelligent entities once there are such? Or is it some other thing?

Expand Your Thinking

1. Try your hand at generating a definition of action. As important as the definition you come up with will be your analysis of the role that a definition is supposed to play in our understanding of what makes action special.

2. Humans are prototypical agents. But not everything we do is explicitly guided by our beliefs, desires, and our evaluations of what behaviors we could perform to connect them. Very few of us say things such as "I have a desire for a delicious meal. I believe that following this recipe here would result in a delicious meal. Further, I believe that in order to follow the recipe, I would need milk, but I believe I don't have milk. So, I desire to get some, and I believe there is some at the store." Moreover, we do not much think about our desires and beliefs and the behaviors that would connect them. We just do stuff. Does that mean we're not acting? Or that the standard theory is not good?

3. You will find a lot in our list of further reading at the end of this chapter, but we highlight two of special interest. First, Sims (2019) pushes back against our idea that we should be doing a lot of theorizing in order to figure out which things in the world are agents because we will know it when we see it. What do you think? Wu (2011) as well as Bello and Bridewell (2017) focus on a theme we have not been able to address: the need for attention in agency. You may end up wondering what attention *is* exactly. So, take some time to reflect on these questions and, if you're interested, read some more on the topics using the resources provided.

Further Reading

Anscombe, G. E. M. *Intention*. Cambridge: Harvard University Press, 2000 [1957].

Bello, Paul, and Will Bridewell. "There Is No Agency without Attention." *AI Magazine* 38, no. 4 (2017): 27–34.

Davidson, Donald. "Actions, Reasons, and Causes." *The Journal of Philosophy* 60, no. 23 (1963): 685–700.

Dennett, Daniel C. *The Intentional Stance*. Boston: MIT Press, 1987.

Dennett, Daniel C. *Freedom Evolves*. New York: Viking, 2003.

Dretske, Fred. *Explaining Behavior: Reasons in a World of Causes.* Boston: MIT Press, 1988.

Frankfurt, Harry G. "The Problem of Action." *American Philosophical Quarterly* 15, no. 2 (1978): 157–162

Marchesi, Serena, Davide Ghiglino, Francesca Ciardo, Jairo Perez-Osorio, Ebru Baykara, and Agnieszka Wykowska. "Do We Adopt the Intentional Stance toward Humanoid Robots?" *Frontiers in Psychology* 10 (2019): 450.

Sims, Andrew. "The Essence of Agency Is Discovered, Not Defined: A Minimal Mind-reading Argument." *Philosophical Studies* 176 (2019): 2011–2028.

Wu, Wayne. "Confronting Many-Many Problems: Attention and Agentive Control." *Noûs* 45, no. 1 (2011): 50–76.

5 A Minimalist Theory of Agency

Introduction

As we saw in the previous chapter, theories of action tend to require agents to have mental representations. The standard story of agency has the belief–desire pair at its center, and entities without this mental stuff do not count as agents. A common trope, in discussions of artificial intelligence that are structured by that view, is that AIs do not have mental representations and so cannot be agents. As we began to see, however, much of the work on agency, although presented as *general*, is best understood as an account of *human* agency. A general account of agency might not need this mental stuff. Here, we provide a minimalist account of agency—one that may be able to accommodate the actions of machines. We suppose that for any system to which it makes sense to attribute agency, there will be representations playing the right regulatory role with respect to the things they are representing.

Don't take our word for it. Let's work through this slowly and see if we can make any progress in challenging the centrality of the belief–desire pair.

I. Beyond the Belief–Desire Pair

Accounts that rely on belief–desire pairs to explain actions seem incompatible with the attribution of agency to the machines we are building because we seem far from having any machines that have desires. Desire is a prototypically emotive state, and emotive states are generally thought of as mental states. Many philosophers are reluctant to attribute such states to any nonliving systems or to systems that are not, at least, composed of other living systems that themselves have such states (e.g., groups of people). Even a belief seems to some as something that only a living thing could have. Machines seem more like they have information, data, or something like that, as do record players and books, but they do not seem to believe anything.

Let's look more closely at beliefs and desires to get a fuller picture of what role they might play in agency. Our beliefs and desires are kinds of representations. Beliefs represent how things are in the world, and desires represent how we would like those things to be. How do these representations come about? Recall the mustard example from the last chapter. What makes me believe that I do not have any mustard, that I would even like it, that there is some over there, and that if I go over there, I can get some?

An empiricist would say that all of these beliefs come into the mind by way of the senses. (Of course, many people are not strict empiricists. But even those people should be willing to endorse a picture that's much like the one to follow. Empiricists just go further and say, "And that's all there is to knowledge," while others think there are other sources.)

Here's the picture empiricists generally give of the source of our beliefs. The world is a certain way right now. I know about it by seeing parts of it around me, by hearing what is going on around me, and also by reading the news or listening to news reports, having conversations with other people, and so on. I reach out and touch things and learn about the world that way. I smell and I also learn something that way. All of this is sensing the world and using those senses to find out about its current state. I can also use this sensory input to find out about the past, about the future, and about the present elsewhere in the world. Maybe I look out the window and see that it's raining, and so I find out that it is raining where I am. And the weather report tells me what the weather is like elsewhere, what it was like yesterday, and even (with variable reliability) what it will be like tomorrow.

This is a standard story, and there is nothing particularly exciting here. But what is sometimes glossed over is just how incredibly complicated the whole thing is from a biomechanical standpoint.

I look out my window and see the rain and trees and grass and horses and whatnot. And I do not make any particular effort to do so. Yet, a great deal is going on. Light waves and sound waves out in the world impinge on parts of my body—my eyes and ears—and various volatile chemicals get into my nose and other parts of the surface of my body come into contact with the surfaces of things. All of those body parts at the surface are connected with other parts deeper inside by means of their own physical processes, generally electrochemical in nature. Somehow or other, all of this activity is able to convey information about the things I am aware of by my senses to my awareness itself of letters on a page, raindrops falling, and voices on the radio. The light, for example, comes from many different raindrops, many different photons from each one, entire sequences of photons one after another. On their own, not much differs between them—a little higher or lower frequency, a little difference in their direction and how they impact the eye. It is extremely complex the things that must be done in order to convert sequences of photons striking the eye into sequences of visual experiences of the sources of those photons.

Now, in the case of people, there is about 3.7 billion years' worth of evolutionary engineering that makes possible the coordination between our beliefs about the world and various physical happenings in the world that convey the information about it. And there is a great deal we do not know about the evolutionary forces that were responsible for our exact way of finding out about the world via our senses. Still, we know enough to know that certain features must be in place and that certain constraints are operative for any system to be suited to learning about the world in roughly this way—that is, by receiving signals from the world in the form of electromagnetic fields (in the case of touch or hearing), radiation (in the case of seeing), chemical reactions (in the case of smelling), and so forth. Finding out about the world requires being connected to it by means information-bearing signals, as well by apparatus that can detect and decode them.

If this view is right, then maybe we do not need the belief–desire pair to explain agency. Perhaps we have found the general function that beliefs, desires, and so forth play in agency: they are representations that guide behaviors. That is going to be our story of agents. They are entities whose behaviors are guided by their representations of the world. All that agents really need is, first, this capacity to represent, and then the capacity to use these representations to control their behavioral repertoires. Agents need to gather information from the outside world, modify their internal states in response to that information, and then make changes in the world prompted by their new internal states. Some of these internal states will be representations of various sorts; some will be causal and structural connections between those representations, including the means by which the changes to other states are made. It is not a story about mental states that's needed but rather one about representation.

We are focused on the question, What do entities need in order to be agents? We are not asking, What do most agents in the world have? Here's our candidate list: representations, behavioral repertoires, connections between various representations, and triggers for behaviors.

In people, these include beliefs (about the world, about what is good, about the way to do things, about others' beliefs, etc.), desires (structural causes of plans and so forth), bodily movements (including vocalizations, facial changes, tone modulation, and eating), and senses (for generating and updating representations). There is also rich feedback between all these elements. In machines, these will include various input–output systems, actuators, feedback control loops, registers in their computer controllers (as representations), and settings and meters (as representations and for generating and updating representations). We are going to try to abstract away from all those particularities and to see through to bare agency.

One of the simplest cases we can think of is the mechanical thermostat. Abstractly, this very simple device has a small behavioral repertoire, a means of representation that connects it to information-bearing signals from the world, and its small behavioral repertoire is guided by which of just three possible states is being represented.

Concretely, this device has a pair of switches that, when engaged, activate the heater or cooler, respectively. The moving part of both of the switches is a bimetallic junction (basically two strips of metal stuck together) that indicates the temperature of the room through its curvature. The state of that junction, its curvature, changes in response to changes in the temperature because its two parts shrink or expand as the temperature goes up or down, but they do so at different rates. That change of state is the thermostat's way of gathering information about the outside world and is also a modification of its internal state. Further changes in temperature induce further changes in that state. When the strip bends enough to come into contact, it closes one or the other switch, and there is yet another change of state: an internal electric circuit is made, electricity flows in it, and the thermostat engages either the heater or cooler, the elements of its behavioral repertoire. One more basic change of state is possible: users may adjust the temperature setting, and that is reflected in a change of the relationship between the curvature of the strip and its point of contact with either the cold or hot actuator. This change is a way of getting information about the world and changing its own state in light of that information. In the same way that the strip itself indicates the temperature and does so by means of the physics of the situation, the temperature setting indicates the desires of the humans using the thermostat and does so by the mechanical relation between the strip and the junctions. That relation encodes the user's desire that the room itself be at a particular temperature.

This simple machine agent has internal representations, one for our desired temperature and one for the current temperature, and a behavioral repertoire that is just sufficient to keep these two representations aligned. If we like, we can think of the *mismatch* between the temperature representation and the desired temperature representation as another representation. That other representation, we might say, is what guides the thermostat's behaviors. You could also simply see the temperature setting and the thermometer as one representational capacity, representing the mismatch between how we want the room and how the room is. The thermostat's behaviors would then be guided by which of the three representations that makes possible: too hot, too cold, just right. Whether we think of this as a single representation or a cluster of them, the key point is that the behaviors are being guided by the representation(s) of interest.

While it may seem stripped down to the bare bones, this minimalist account seems to us to strike just the right balance between making sense of things we already think of as agents and actions (us and the things we do, cats and dogs and the things they do, and so on) and allowing us to make sense of other systems as agents with their own actions (maybe certain kinds of AI, maybe certain kinds of robots, and also maybe various kinds of control mechanisms, such as machines that deploy other machines given the right kind of circumstance, for example).

II. The Possibility of Machine Agency

Now we think you can see why we suggested earlier that looking for the right analogues of various things that humans do when they act might be a mistake. It's a mistake in part because thinking of representation as a "mental" phenomenon obfuscates how representation works. Consider the following line of reasoning, which, while perhaps not behind all cases of reluctance to grant agency to machines, is salient in discussions of machine agency. When we explain action, rather than mere behavior, using the belief–desire pair model, we notice that both members of the pair, as well as their connection, require representations. First is a representation about how things are in the world, second is a representation about how things ought to be in the world, and third is a representation of what is needed to close the gap between the representation of the world as it is and as it ought to be. Because all of these things (beliefs, desires, and representations) are mental, then we can conclude that mental states are required for anything that counts as genuine agency. Again, no machines we have or are likely to have in the near term have mental states. So, no machine acts.

It should now be easy to see that this line of reasoning rests on two common misconceptions—at least, we think they are misconceptions. The first is the supposition that the belief–desire pair is doing the explanatory work in the story about action above. The second is the supposition that representations are mental. Our working account of acting and of agency avoids these. To clear up the first misconception, we have been supposing that the explanation for agents acting in the world should be detached from belief and desire as such. Instead, we think we can make do with representation and then understand goal seeking as behaving *for the sake of* (behaving in ways that are *directed toward* or *guided by*) a representation of the world that indicates that the world is within acceptable parameters. For example, I continue to sweep the floor until my visual impression of things (a representation of the floor's state of cleanliness) is sufficiently close to a specified end state of things (the floor being impeccably clean). Likewise, the thermostat continues to energize the cooler or healer until its representation of the current temperature is "just right".

Beliefs and desires are red herrings on the hunt for a *general* account of agency. Entities like us (humans) do apparently have such states, and we are motivated by them to act to bring our representations of the world in line with our representations of how it ought to be. The causal story behind that motivational component of our action is a wonderfully and massively complicated account of the evolutionary history of our minds—an account that is only partly written. But it's an account of the history of how *we and creatures like us* are motivated to act rather than a general account of agency. We think that any system that can be said to be behaving for the sake of some end suffices as an agent. Having this kind of general account allows us to make sense of entities that move through the world and act in ways that are distinctly unhuman-like.

Note that this account allows for agency to be exercised more and less *strongly*—that is, some agents will have more or less agency in their actions, in their selection of behaviors for executing those actions, and for setting the end states that those actions are directed toward. Not all agents look or act the same, and there are degrees to which agency can be exercised. More sophisticated agents will have more and more interesting representations and behaviors in their repertoires. Compare, for example, what kind of action is available to a thermostat versus a self-driving car. One of the most important ways of expanding the capacities of machine agents is to make them more sophisticated in their means of taking in and representing information. The thermostat performs the same simple routine over and over: strip bends, power on; strip unbends, power off. One could imagine writing a flow chart for this activity: if the strip bends left too far, turn on the heater; if it turns right too far, turn on the cooler. Described that way, we can think of what the thermostat is doing as executing a very simple algorithm. A good way to get insight into what more sophisticated agents do is to think of them as executing ever more complex and powerful algorithms.

In summary, we have noted that beliefs, desires, and intentions are all ways certain representational capacities that are important for agency are implemented in humans, and we have also noted that this way of doing things is the result of our own evolutionary heritage. While it may possibly be that any real agent must have beliefs, desires, and intentions and all the things that we have that guide our behaviors, it may well be that such things are *only* necessary for human (or mammal, animal, or what have you) agency and not for agency in general. What *does* seem necessary, on our view, is that, in agency, behaviors are guided by representations.

III. Critiques and Objections

We have proposed, and offered reasons to accept, a particular account of agency—one that treats as agents any systems that exhibit behaviors directed toward minimizing the divergence between (at least) two separate representations (in short, behaviors guided by representations). As minimalist as it is, we have a fully specified theory that in addition to what we see as its theoretical virtues has another nice feature: it makes appropriate contact with theories of human agency without using human specific tools. First, our account self-consciously adopts Frankfurt's guidance condition for action. We can also see reasons-based accounts, like the belief-desire pair theory of human action explanations, as instances of our *behavior for the sake of representations* account. Second, the theory is compatible with various kinds of functionalist accounts, including Dennett's intentional stance, though it has the extra virtue of displaying the objective grounds necessary for adopting that stance in particular cases.

We like this account, and we think it is pretty good. But notice that all the other accounts we looked at came with reasons to accept them, and their authors all thought they were pretty good, too. However charitably you have read this account, many of you, and perhaps your instructors as well, will already have seen things that you do not like, and you may even have an account of your own that you think is much better. So, how can you be expected to go on? That question is itself an occasion for another lesson from philosophy. In short, the lesson is that concepts (and their definitions) are our tools, we are not theirs. We are going to be using the account of agency that we do because it will be helpful for thinking through questions about machine agents. It will not be necessary for this task that we have the final, best, unbeatable account of agency. Instead, we will need an account that is serviceable for its end, and this one is eminently so. Remember, we are not here to sell you just the right story of agency to go put on a shelf somewhere. Instead, we are trying to interest you in sharing with us an exploration of some of the conceptual terrain.

We are going to point out a few ways you might want to critique the theory, even as (we hope) you go on using it. What might one think is missing from a minimalist theory of agency? What might be wrong with it? This is a useful philosophical practice of its own: anticipate others' objections. That serves at least two worthwhile ends. First, it makes you more sensitive to and aware of how others will be taking up what you say. That will improve your communication skills. If you are really speaking *to* the folks you are addressing, then they are more likely to understand you. Second, it makes you more sensitive to the limitations of your own view. That will give you an insight into how to modify it to make it hew closer to the truth of things, which is, after all, what most of us are seeking in philosophy.

Before we offer some ways you might want to critique things, we will respond to one objection by way of clarification of the view. Incidentally, this may give you a sense for how to respond to objections. Many people have argued that there are simply not enough explanatory resources in a view with only behaviors and representations guiding them. It is often supposed that a minimalist account of representation such as ours cannot work because it is inadequate to the rich networks of meaning that attach to human-level mental representations, especially including intentions. But this understanding of the situation is badly inverted.

While it is true that the sophistication of human intentional states does allow very simple "behaviors for the sake of" to be understood purely in terms of the relation between the agent's mental states and the part of the world of which its representations are representations, the capacity to create and sustain such a connection between representation and behavior is held only in conjunction with other agents in a meaning-sharing context. In part, this is because various behavior–representation pairs only constitute the actions they do in such a context. Generally, the weight given to the representation as constituting the act is balanced against the weight of the context itself, and part of being a human fully inculcated into a social practice is to know how to apportion these weights. Indeed, it is a standard notion from action theory that kinds of behavior depend crucially on context in order to be the act that they are. A simple example is analysis of blows from a closed fist. By turns, these are punches in a boxing match, assaults, elements of a stage play, and so on. Context is important for making them the acts that they are.

We contend that what is missing from efforts to understand autonomous systems as agents is not that they don't have representations and not that representations need to be richly structured in order to contribute properly to agency. Rather, it is that our focus has not been expanded to include the practices that embed those representations that they do have into contexts where action ascriptions are appropriate. Our general account of representations and their roles illustrates and relies on the fact that representations together with context can suffice to generate agency without the need for beliefs or desires.

Now let's turn to identifying some places where it might be more fruitful to object to the theory. The things that might be *wrong* involve what it might entail about the role of representation in marking the difference between action and mere behavior. While we talked a lot about minds before, we have not had much to say about consciousness, for example. Are they the same? Maybe minds are not particularly important for the concept of agency as such. But it seems far from our own lived experience of agency to think that consciousness is not. At least being *conscious* does appear to be closely connected to what we think of as the difference between our bodies simply doing stuff and us *acting*. That is a persistent and powerful worry.

Another possibility for what is wrong with the view is that even in humans, one might think, beliefs, desires, and so forth (the representations we use for acting) are only our own ways of *explaining* our experience of agency. As such, they do not have anything fundamental to do with agency itself. So, we might be chasing a red herring. Agency, one might think, is deeply buried in the structure of the biological systems that we are, and our thoughts and feelings are ways of getting insight into that but are not the main story.

This is a good point to consider. But it changes the focus from explanations to the nature of things in themselves (what philosophers call "ontology"). A theory of action *explanation* might require that we get the phenomenal features right but not that we get right the deep nature of things. (This is similar to the way physicists think about there being laws of physics at our level of observation that are correct independently of whether we have the right story about the underlying physics that gives rise to them.) This would be a problem for anyone with an account of the agency of machines in advance of an entire evolutionary history with these machines. It would be hard to move forward from explanation to ontology if we could not rely on the phenomenal adequacy of representations coupled to behaviors as a clue to the fundamental story of agency as such.

Here's a third problem. One might claim that while agency is more or less what we think it is in the case of people, and even smart animals, the fact that the beliefs, desires, and so forth that separate our actions from mere behaviors are representations is not really the crucial thing. Instead, it is the precise *kind* of representation or the precise *kind* of mind that has them or something of that sort that is important in distinguishing acting from behaving.

The considerations above might prompt you to think a little more about what really does separate behavior from action. Why does one care about the difference in the first place? There are at least two reasons. First, there seems to be something special about the things that we do when we're most fully human—making moral decisions, say, or caring for others—that seems different from most of the other things we do, and that most of the other animals do, and those doings seem special somehow. So, we want to understand what is special about them. Another reason is that sometimes when people do things, we think it is appropriate to hold them responsible for those things, and we might praise them or blame them, or do any of the other things that we do when folk are responsible for things. But what separates those doings from the doings of the weather, of animals, of plants, and so forth? And if you think that people, as interesting and complicated as they are, are also at the bottom of physical systems that obey physical law like all other physical systems, you might be puzzled about what it is that separates their doings from the doings of the rest of the world in some instances. We will keep exploring these difficult questions in the chapters to come.

Those are a few kinds of problems. Can you think of more? Are they problems that can be solved now, or would we have to wait for more, perhaps much more, experience with machine agents? A really good way to object to someone's argument is to think of examples that most people will already know about, that are easy to analyze, and at least seem to conflict directly with the claims you are objecting to. It is less effective to think of obscure, complicated cases that are themselves controversial. The main thing, as always, is to make sure that you are responding to what your interlocutor is actually saying, so that you are focused either on the factual aspects of the premises being used or on the connection between those premises and the conclusion being drawn.

Rather than considering further objections, we are going to leave aside explicit consideration of the theory of agency. Our task now will be to try to understand how it is that the machines we have built (and might build in the future) could even be agents according to our account, and also, what that would look like from a conceptual point of view. What will be coming is a gentle introduction to the way that machines can deploy elements from a behavioral repertoire because of various representations that they have. We will see that a great deal of the sophisticated behaviors and their associated actions that seem characteristic of humans can be accomplished by these machines using resources that are not conceptually very far from the simple representations and limited behavioral repertoire of the thermostat.

Your Tasks

Test Your Understanding
1. Generally speaking, why do philosophers who endorse the belief–desire pair resist attributing agency to machines?

2. How do we define "representation" as presented in this chapter? What is the difference between a "representation" as defined in this chapter and "mental representation"?

3. What does it mean for an account to be general and minimalist?

Reflect or Discuss

1. How would the philosophers we discussed in the previous chapter respond to the minimalist account? What could they say? Which objections would they raise? Might some of them find the account satisfactory?

2. Recall the views proposed by Searle and Turing. Which of the thinkers might be more sympathetic to the minimalist view? Why?

3. What if some theory of agency came to what you believe to be the wrong conclusion and classified something as an agent that you think it should not? What, philosophically, might be the right response?

Expand Your Thinking

1. According to the minimalist theory presented in this chapter, are LLMs a kind of agent? Why or why not? If you need more information on how LLMs function, do some digging.

2. Does the acting versus merely behaving distinction matter for questions of responsibility? Why or why not? Can you think of examples where it does matter?

3. Are the agential capacities of an entity, whatever kind of entity it may be, relevant for the kinds of relationships we might form with the entity? Or is agency a red herring in this context?

Further Reading

Barandiaran, Xabier E., Ezequiel Di Paolo, and Marieke Rohde. "Defining Agency: Individuality, Normativity, Asymmetry, and Spatio-Temporality in Action." *Adaptive Behavior* 17, no. 5 (2009): 367–386.

Bello, Paul, and Will Bridewell. "There Is No Agency without Attention." *AI Magazine* 38, no. 4 (2017): 27–34.

Callender, Craig, and Jonathan Cohen. "There Is No Special Problem about Scientific Representation." *Theoria: An International Journal for Theory, History and Foundations of Science* 21, no. 1 (2006): 67–85.

Lakatos, Imré, and Alan Musgrave. *Criticism and the Growth of Knowledge*. Cambridge: Cambridge University Press, 1970.

McDowell, John. "Pragmatism in Intention-in-Action." In *New Perspectives on Pragmatism and Analytic Philosophy*, edited by Rosa M. Calcaterra, 119–128. Leiden: Brill, 2011.

Popper, Karl. *Conjectures and Refutations*. London: Routledge, 2002 [1963].

Winther, Rasmus Grønfeldt. "The Structure of Scientific Theories." In *The Stanford Encyclopedia of Philosophy*, edited by Edward N. Zalta and Uri Nodelman. Stanford: Metaphysics Research Lab, Philosophy Department, Stanford University, Spring 2021. https://plato-stanford-edu.proxy.library.georgetown.edu/archives/spr2021/entries /structure-scientific-theories/.

6 Computational Implementation of Agency

Introduction

We will now continue our discussion of the minimalist account of agency, but this time with explicit connection to its implementation in machines. We are going to see how machines, once they have been given all the necessary elements of agency, can have that agency augmented. This is done by expanding their number and type of representations, their behavioral repertoires, and the connections between them. Everything here other than the substrate applies to agents in general, but the discussion is framed around computing machines and machines guided by and integrated with them. The idea is to see how to implement agency in these built systems—agency that extends the simple agency of a thermostat in the direction of more sophisticated agency, one might say.

Our minimalist account grounds agency in representations and some kind of behavioral repertoire. In people, and other evolved agents, their capacities to represent (feelings, and intentions, and minds) and to behave (run, and skip, and hop, and jump as well as shake hands, and talk, and so on) grow along with them. Machines will generally have these things built into them either as they themselves are built or later when modified. Even so, however, it can be a challenge to see where the representations are, which things are behaviors, and how they fit together. In the case of the thermostat, it was trivial to point out the bimetallic strip as a representation of the current temperature and the locations of the actuator junctions as the representations of the coldest and hottest it should be, respectively, and finally the connection between the strip and a junction as the trigger for the deployment either of the cooler or the heater. It is also trivial to identify the simple behavioral repertoire of actually turning on and off the cooler and the heater. It is not too difficult to see how, according to the minimalist account, agency is implemented in that device.

For more sophisticated systems, it is not so clear what is going on, and it is not even clear on initial inspection that things such as self-driving cars and strategically sophisticated gameplayers do implement agency in anything like the way the thermostat does, much less in a way that is closer to how we do it. Yet, if they are agents, then they need to represent and they need to behave, and that behavior needs to be guided by and directed toward those representations. How do they do it and how is that related to what the thermostat does? The *short* answer is that they consume information-bearing signals, perform computations on them, and then generate more information-bearing signals. But what does that mean, and how can that be all there is to it? Where in all that were the representations, and where in all that was their behavior directed toward them? That will require the *long* answer.

We will get to the long answer, but for it to make sense, we should start at the beginning, with algorithms and what it means to compute.

I.   Algorithms and Computation

One hears a great deal about algorithms today. We hear about them in the context of social media platforms, news feeds, online shopping, and, increasingly, in policing, finance, and healthcare. But we hear less about what algorithms are and how they work.

In the first place, while algorithms are commonly associated with electronic computers, their history is much longer. (See Pasquinelli 2019 for an informative but idiosyncratic rendition of that history.) There is a long tradition of using the word "algorithm" to denote a structured, ordered process where the next stage of the process is contingent on the current stage. What computers do when they compute is execute algorithms of one sort or another. What machines do when they control other machines is execute algorithms of one sort or another. When phones run apps, when televisions render images, when digital cameras generate images from light waves, they are all executing algorithms. Similar considerations apply to people tying shoes, doing long division, and playing tic-tac-toe. Indeed, bird murmuration, the massive collective motions of some bird flocks, is itself the execution of an algorithm.

While often presented as a series of formal rules, this notion of a structured process can also apply to more or less intuitive instruction sets, and instruction sets that require those implementing them to be aware of background considerations. For example, if you and I are both experienced chefs, I can give you a recipe for some dish that is very informative to you that gives you all the information you need at each stage to complete it and move on to the next, by appealing to the fact that I know you know things about, say, poaching or the proper proportion of butter to flour in a cream sauce. That set of instructions would count as an algorithm, but only against the background of your basic skills counting as primitive elements of the process. There is no compelling reason to reserve the word only for contexts appropriate to computers.

In philosophy, we often use normal words in very precise ways, and that can be confusing (and annoying, if we're honest). It is done to be sure that any conclusions we draw about a subject come from things we know are features of that subject, not an accident of casual word choice. While there is no reason to reserve the word only for contexts appropriate to computers, still, the role of algorithms and algorithm design in the context of computer science is difficult to overstate. There is utility in clarifying some added constraints that are required in the case of algorithms appropriate to those contexts. Algorithms are the result of breaking a large or complicated task into less complicated or smaller subtasks so that, by repeatedly performing one or the other of the subtasks, eventually the large or complicated task will be done. Even something as simple as moving a pile of rocks yields an algorithmic solution: if there are any rocks in the original pile, pick up the nearest one and move it to the new pile; if not, you're done. Here, the entire pile is moved one rock at a time. While you never "move the whole pile of rocks," the whole pile is moved. That's a silly algorithm to instruct a human to perform. We know very well how to move piles of rocks, but computers do not until we tell them. It is interesting to note that the elements of the rock-moving algorithm are, in various combinations, basically all that is needed to tell any computer how to do anything it can be told how to do.

Algorithms are defined by the processes that they spell out, not merely the task they accomplish. Some algorithms are better suited to accomplish a task than others that accomplish the same task. "Better suited" can be a matter of being more convenient for the user, having more easily accomplished subtasks, having fewer necessary steps, or something along those lines. For algorithms that use the same basic subtasks, there is still a ranking of them into the more or less "clever" as they need more or fewer steps (on average) to accomplish the same end. This is nicely illustrated by sorting algorithms, of which there are many. We will illustrate the point by considering just two.

Suppose I want to alphabetize the books on my shelves. There's the way I do it, the naive way. That is, grab any old random book and make it first. Then, take the next book I see and put it either to the right or left as needed. How might one do that? Well, compare the first letters of the authors' last names. Put those with letters earlier in the alphabet to the left of the other. If they are the same, do that for the next letter, and so on. If the names are the same, turn to the titles. Then, do that for the next book with respect to each of the books already there. That's very costly and time-consuming, but it does not matter for a few books, and I am also doing most of the work as I pick up the individual books off the floor. So, I do not notice the cost—that is, because the pile is small, I can perform the costly computation of figuring out where a book goes as I'm picking it up and standing up and so forth. But that only works for small piles. A better way in general is merge sort. Merge sort tells us (in this context) to take the first two books in the pile and alphabetize them, then the next two, and so forth. If it's an odd-numbered pile, we just leave the last one alone. Then, we go back to the beginning and take the first and second pair, merge them together alphabetically, then the next pair of pairs, and on we go. Eventually, the entire pile of books is alphabetized. The average number of things one has to do in the naive process is a lot more than one has to do in the merge-sort process: n^2 versus $n lnn$. For 10 books, that is 100 versus 23, for 100 books (still a very small personal library), it is 10,000 versus 460. For the 51 million or so books in the Library of Congress, it is 2.6×10^{14} versus 9×10^8. You can see that if there are many objects to sort, merge sort is a significant time-saver.

Not every step-by-step process *is* algorithmic. Indeed, the dividing line between the things that are algorithmic and those that are not marks another important distinction. It used to be seen as the distinction between living and nonliving systems. A rock rolling down a hill under the force of gravity is not algorithmic. A hurricane, even though it is self-sustaining and self-organizing to some extent, is not algorithmic. On the other hand, arguably, the building of a cell according to instructions in DNA is the execution of an algorithm. Generally, living systems involve the execution of various algorithms. Now, it has become common for nonliving systems, computers especially, to execute them as well.

Much of what we humans do in getting along in the world is executing various kinds of algorithms. Here's a simple example: Humans are good at adding and subtracting, multiplying and dividing, even though we are slow at it and sometimes make mistakes. The way we do these things generally involves two distinct processes. The first is that most of us have internalized some very basic operations. Without being aware of what we are doing exactly, most of us can multiply any pair of one-digit numbers and we can divide, say, a two-digit number by a one-digit number and note the remainder. Similarly, we can add and subtract one-digit numbers without being aware of what we're doing. Try dividing 10 by 2 or subtracting 4 from 7. Most of us do most such operations very quickly. It's just a bunch of basic stuff that we must learn by rote at an early age. But when it comes to much larger numbers, most of us cannot simply know the answer without working it out. That's where the second process comes into play.

We are trained in a variety of algorithms that allow us to leverage our basic skills into the capacity to add, subtract, multiply, and divide any numbers at all, no matter how big. Let's start small with 13 multiplied by 24. The exact way we do that depends crucially on the way we represent numbers as strings of numerals from 0 to 9. For the case of multiplication of two numbers represented as strings of numerals representing first the count of 1s, then the count of 10s, then the count of 100s, and so forth, we begin with the last element of one of these strings (it could also be the first, but most of us are taught to begin with the last). In our case of 24, we take 4. Then, we multiply the number represented by that numeral first by the number represented by the last numeral of the other string, the 3 at the end of the string 13. Then, there is a decision point: if that multiplication results in a number greater than 10, the numeral representing the count of 10s in that product is "carried" over to the next element of the second number to be multiplied. In our case, the number is 12. So, we carry the 1. Now, things get tedious, but it's also worth seeing explicitly. So, just bear with us for another couple of paragraphs.

What does "carry" mean here? And what do we do with the numeral in the 1s place—the 2 in 12? First, we take that numeral and make it represent the 1s place of a brand-new number we are creating. To carry, we just keep note of the numeral 1 until the next stage of the process. How does one do that? Well, we stick with the 1s numeral of the first number and now multiply the number it represents by the number represented by the numeral in the 10s place of the second number—so, 4 multiplied by 1. Having done so, we add in the number represented by the "carried" numeral, and that gives us a 5. Now, we take the numeral in that number's 1s place and put it in the 10s place of the number we started making in the last step, and we carry the numeral in the 10s place to the next element of the second number to be multiplied.

In our example, we already finished this stage, but typically we go on multiplying from element to element and continue this until there are no more elements in the string representing the second number. If there is a carried numeral at that point, we stick it in the next place (10, 100, 1,000, or what have you) of the number we are creating. If there are more numerals in the string representing the first number we are multiplying, we make a new number. But the string that represents this number will start with a 0 in the 1s place if we're in the second round, and if we are in stage number n, it will have a 0 in the first $n-1$ places. Then, we repeat the number-creation process described above, but now using the numeral in the string representing the first number's *next* slot and beginning with putting the 1s result of the first multiplication of the second number's representation in slot n of the current representation of our new number. So, for us, it's a 2, and we multiply that by 3, getting 6. So, with nothing to carry, we start a new number down below with 0 in the 1s place and 6 in the 10s place. Then, 2 multiplied by 1 gives us 2, and we put that in front of the 6, in the 100s place. We continue *this* until we are out of numerals in the first number representation. At that point, we will have as many new numbers as the first number representation had digits. For us, they are 52 and 260. We add all of those together, and we are done. How do we add those big numbers together? Well . . . there's an algorithm for that.

Forgive us if your eyes have glazed over at this point. This is all very complicated in one respect and very simple in another. All of us know how to do it, but it is not so easy to describe it without implicitly using the knowledge of the process the person it is being described to already has. (And probably we did some of that even in our very long explanation.) But how could we get a machine to implement this algorithm when it does not have any knowledge of the process or any knowledge at all? There are two kinds of ways of doing this. The first is purely mechanical and, as such, limited to numbers of a certain size that is determined by the construction of the machine. Blaise Pascal took early steps in this direction, making a mechanical adding machine, and then Gottfried Wilhelm Leibniz invented a machine that could multiply numbers together. These machines were able to use gears to advance counters and keep track of the various stages of the process, including the numerals to be carried at any stage. Pascal's adder was operational, but apparently Leibniz's "stepped reckoner," as it was called, was never fully operational, although the principle is thought to have been sound. These machines operated only on number representations of fixed length—eight or twelve digits in the case of Leibniz's machine. That's fine if you never need to multiply or add any larger numbers. You could, of course, just keep building larger and larger adding and multiplying machines, but that seems costly and inefficient unless adding and multiplying are the only algorithms you ever want a machine to perform. Otherwise, you would have to build not only bigger and bigger machines but also special-purpose machines for any algorithm you might want to implement. But there are many different algorithms one might want to implement in a machine.

For example, you might want to weave very complicated patterns in fabric. This is the kind of thing that takes a lot of skill, expertise, and concentration for a person (or team of people) to do on a loom, a machine for weaving fabric. But what if it could be mechanized? Moreover, what if the same loom, on different occasions, could be used to make very different patterns? This is what the Jacquard loom head allows. It provides an algorithm that needs to be followed as the various threads are woven in among each other in order to result in a given pattern, and it provides the machinery to implement that algorithm in the rest of the loom. (See Essinger 2007.) It works by means of stacks of perforated cards that can be changed out in favor of others that give the algorithm for other patterns. These are simply cards that either have or do not have holes in regimented locations, and the presence or absence of a hole either allows or prevents a threaded needle from passing through. The card stacks are, in effect, algorithms that specify the weaving pattern. So, rather than having a single machine for each pattern that you might want, there can be just one machine that implements many different algorithms.

Charles Babbage applied this flexibility to the case of mathematics. His analytical engine was a machine that could perform a variety of different calculations, depending on which algorithms were "programmed" into it.

By and large, this led to the programmable computers that we have today. Although our computers are mostly electronic and Babbage's analytical engine was mechanical, they occupy the same side of the dividing line between special and general-purpose computers, between those that cannot and those that can be programmed. Babbage's programmable computer was a big step, but programmable computers without programs are like electric cars without batteries. Ada Byron, Lady Lovelace, is widely credited with providing the first of these programs. (Strictly speaking, one could not program the computer because there was no precise outline of how the punch cards would work for the analytical engine, and the punch cards *are* the program. She provided, instead, the algorithmic structure one would need in order to generate that program.) Indeed, Lovelace seems to have understood earlier than anyone not only that these programmable computers could perform all manner of different operations on numbers, but also that they could be further generalized to music composition, image generation, and so forth (Hollings et al. 2018).

In computers, algorithms can be more or less clever. It is the nature of the sequence of steps and the costs to perform those steps that tell us how clever an algorithm is. More efficient to the task, fewer steps, easier steps—these all go into the assessment of the cleverness of the algorithm. The standard way to calculate cost is to evaluate how many steps it takes (on average, in the best case, in the worst case, and in the typical case) to perform a given task on a certain size of input. This has its limitations, since certain algorithms may be objectively more costly than others but well suited to the task at hand, given the expected size of input and initial costs of getting the program running in the first place.

We typically speak in terms of computing the values of a function for a given input. The most typical functions we consider are those that take in one or more counting numbers and output a single counting number—like addition, or multiplication, or the successor. To compute a value of a function is simply to take some input and give back the value of the function on the input, subject to various constraints. What are those? Well, this is where it gets vague, but the idea is that the output must result from some definite procedure that, step by step, without any insight or discernment, transforms input to output. These days, those constraints are taken to be tantamount to being something a machine can be programmed to do, although in the past, computing was, in fact, a job that a person had. Any task that required adding pages of numbers, or calculating trigonometric tables, or figuring out the orbits of planets was a job for a calculator—a person who did those calculations. Even then, though, the idea was that in the parts that mattered, in the parts where the computing actually got done, there were no decisions to be made—computations would always be performed the same way and were thus subject to check. If more than this were involved, there would be ambiguity about whether the same functions were being computed in each case.

A Turing machine is simply one model of what it is to be a computer. With this model, we write ones or zeros on a long tape, moving back and forth following instructions, never knowing where we are on the tape or what stage of the computation we are at. Yet, Turing machines can compute anything that is known to be computable, and according to the Church–Turing thesis, anything that *could be* computable. The Church–Turing thesis is the thesis that there is only one kind of computation—that is to say, if you come up with a brand-new kind of computing, even quantum computing, there is nothing you can compute with that new method that cannot be computed with the methods we already have. While this thesis is unproven, in part because there is no completely general acceptable definition of computation, it is widely accepted as true by mathematicians, logicians, computer scientists, scientifically minded philosophers, and many others. Moreover, every single explicit definition of some model of computation that has been proposed has also been shown to be equivalent to all the others. These are good reasons for us to accept the thesis. If true, it has an important consequence: to understand *completely* one model of computing provides understanding of all of them. It does *not*, of course, provide any skill with using all or even that one, but it does provide a conceptual grounding of the notion of computing.

There are no physical Turing machines. Turing machines are abstract in the same way that numbers are. Even so, our contemporary electronic computers are remarkable and powerful devices, and while they do not have unlimited instruction sets or arbitrarily large storage capacity, they still do a good job of approximating Turing machines, at least for the computations that we hope to perform. On the one hand, they have intrinsic limitations, as do all implementations of formal mathematical systems. (Arguably, humans also suffer from those limitations, although we will not try to convince you of that here.) More urgent, though, for their capacities as agents, is that computation is not really overt behavior of the sort that changes the world in recognizable ways. To be sure, physical computation is physical behavior even if normally confined to the interior of the system doing the computing. But still, we think of agents as doing more than computing (or thinking, if they are people): they use their computations (thoughts, for us) to regulate their behaviors in the right way. And Turing machines do not do that, nor do any of the models of computation we know about or that we think are possible. So, where does that leave us?

II. Algorithms, Representation, and Guiding Behavior by Representations

To function as agents, our computers must be embedded in broader physical systems—systems that display overt behaviors and do so in light of the way the world is and directed toward changing it or keeping it from changing. Still, what is necessary to control the behavioral repertoire of those broader physical systems is what is provided by the computers they embed. Recall what we learned earlier: a representation is a system that stands in for another system. Generally, for physical systems, it will be the physical states of the one standing in for the physical states of another. But sometimes, as with numerals, the thing itself stands in for another thing: a number (either an entire number, or for the number of ones, or tens, or hundreds, or what have you, that that number's digital representation has in it). The machines we are focusing on as potentially becoming ever more agential are machines that are controlled by digital computers of various sorts. How do machines of *that* sort represent?

It's tricky to understand just what is and is not a representation in a machine. For example, it is tempting to think that a calculator represents various numbers and mathematical operations as it carries out its calculations and leave it at that. But that will not do. First, the images we see in the calculator's display are used *by us* as representations of numbers and operations. But the calculator is mirroring one of its internal states there, and, at most, we could say it is representing that state by the image on the display. What does this mean? Well, the calculator does not *really* work with numbers. It simply transforms the states of various transistors according to the settings of other transistors. During these transformations, there is nothing, in general, in the machine that is a representation of a number or of an arithmetical operation. Instead, it works directly on the state of the world in front of it and transforms that state in accordance with its built-in structure. That's the nature of digital computation. When we use basic computers such as calculators, we *use them* to represent numbers and mathematical operations, but they themselves do not represent much of anything. Their behaviors are being guided not by systems that stand in for other systems but rather by those systems directly.

Now, think of computers that are harnessed to sequences of behaviors, such as a self-driving car. The behavior of stopping is activated by the mismatch between a representation of the location of a pedestrian, say, and the representation of the safe distance between the car and pedestrians. The *component* computations that produce a representation of a pedestrian in the danger zone are themselves fantastically complicated and necessary for that production, but they are not representational. Again, it can be very tricky to evaluate whether something is representing in a way relevant to the guidance of behavior, but that is what separates agency from everything else. So, it is worth working at.

On the view of agency built around behaviors guided by and directed toward representations, the very heart of what we do as agents getting around in the world is computation. Representations are connected to the world, to each other, and to behaviors computationally. Does thinking that this view of agency applies to us mean that we must think of ourselves as no more than computers, carrying out inexorable programs without any freedom? No, not necessarily. But part of what it *does* mean, from this perspective, is that what computers do is the same kind of thing that is part of what makes it possible for us to be the agents that we are. In ways both conscious and deliberate or inadvertently and offhand, the engine that drives not only the decision-making part of our behavior but also the execution part is a species of computing.

Consider the following. We plan our days by sorting out what tasks must follow others, which are most important, and which are most urgent. We decide whether to put on the brakes or hit the gas when we see a yellow light by tallying up the distance to the other side of the intersection, figuring the time between yellow and red in our direction (and, if we are risk-takers, between yellow in our direction and green in the other), and sorting that all through the current speed and acceleration (and decelerative) capacities of our vehicles. We locate objects in space in part by performing what amounts to a fast Fourier transform of the frequencies of light waves hitting our retinas. All in all, we are deeply computational beings. And we have not even spoken about how we go about unraveling our tangled social threads and joining them up into nice patterns.

Not only computers and self-driving cars and us, but even the thermostat computes, albeit a very simple function. The thermostat compares its representation of the room's temperature to its representation of the goal temperature—that is, if the goal temperature representation is T_0 and the current state of its temperature representation is T, then it computes their inequality. Its behaviors are guided by the representation of the result of that computation in the following way: if $T < T_0$, then heat; if $T > T_0$, then cool. The result of the computation is represented by whether one or the other or neither switch is closed.

But is it really computing these things, or is this us characterizing it that way? The machine itself just goes on and off in the right circumstances. While that is right, that is true for *all* computations. For us, meaning is important, and it structures most of our own agency in the world. But the computations are still being performed after all. What matters for agents *generally*, on this account of agency, is the connection between the computation and the activation of the thermostat's behavioral repertoire.

Even so, while computing is necessary, it is not sufficient for agency. Indeed, there are many computational systems that are non-agential. We can, though, say something about how computation can be used to transform certain systems into agents. What computing allows for is the comparison between two states of a system—states that represent the world and acceptable end states and even the mismatch between representing states, for example, and for the transformation of one or the other of those states in light of changes in another system's state. The algorithms, the step-by-step processes that computers implement, can, in the right circumstances, activate behaviors of the machines of which they are a part and then, in the right circumstances, deactivate them and activate others. That is all to say that systems that compute when appropriately linked with systems that perform various worldly behaviors can act in the world.

Of course, not everyone agrees with this. On the accounts we saw earlier, intentions, desires, beliefs, and other kinds of mental states that humans have are necessary for action. We can now respond more clearly to one of those challenges posed by Searle. Let's take a look back at the so-called Chinese Room thought experiment.

III. Program Implementation Clarified

In his thought experiment, Searle tries to convince us that computers could never be intelligent, in part by arguing that computers are merely implementing a program. One way to respond to Searle is to say that he is wrong in thinking there is any such thing as *merely* implementing a computer program.

Any implementation of a computer programs requires a specific machine with specific architecture on which it will run. Every machine that is implementing a computer program comes with an array of causal powers that are necessary for that implementation. Stated that way, it is unclear whether Searle's argument is right or wrong, and indeed it is hard to adjudicate this debate at all, given how ambiguous many of the terms are. But it is worth seeing why he is wrong to think that implementation is at the level of syntax, in part because it is rarely pointed out in discussions of his thought experiment and in part because we do think that even if his *argument* that pure syntax can never bootstrap into semantics is correct, his conclusion that computers cannot think, have mental states, or be agents does not follow from *that* argument because computers do not work at the level of syntax.

Computers are *said* to operate on syntactic elements in executing their programs. Indeed, the mathematical theory of computation is often framed in just this way: meaningless symbols are manipulated according to static rules, and then the meaningless output is interpreted by humans as the result of a computation in arithmetic, say. But this is a confusion, and it is one part of why thinkers have generally spoken past each other in their analysis of Searle's so-called Chinese Room thought experiment. While the thought experiment is widely critiqued, it is also widely endorsed, and while in practice it is easy enough to get proponents of one view or the other to say they understand what is going on in the thought experiment and what the results will be, it is exceedingly difficult to get them to agree on the significance of the experiment. Even the standard critiques of the experiment uncritically accept the claim that standard computational systems operate on syntactic elements and thus do not represent the targets of their computations. But this, again, is a confusion. No *physical* computer operates on syntactic elements. Instead, they operate on physical elements that are themselves bound up in various representational practices. *Abstract* computers are mathematical objects that do not display any behaviors whatsoever. As such, they cannot *implement* programs. Concrete, embodied computers are by necessity bound up in the practices of those who deploy them and are affected by that deployment.

Turing machines are the basic model of computation around which all our own computers are built. They can be understood as nothing more than specifications of how to manipulate long strings of ones and zeros on an arbitrarily long tape. Those specifications make no reference to what is on the tape already or which part of it is being worked on. Yet, Turing machines can, as far as anyone knows, compute anything that can be computed. They can be incredibly complicated. Some, however, are very simple indeed, such as one that would add two numbers together. But even so simple a thing as that is only the computer that it is in the context of a specification of appropriate initial conditions: that there be exactly two blocks of ones on the otherwise blank tape. Were there fewer blocks or more, it would not be executing the two-digit addition program when it went to work on the tape. It is a standard exercise in classes on the abstract foundations of computing to see what would happen were various Turing machines to be set loose on nonstandard inputs. Very often, nothing happens, but sometimes, the machine will compute something new and unexpected.

Far from showing that Turing machines, being merely syntactic, are inadequate to represent numbers, this serves merely to show, on our view, that like all such representations, the contribution from outside context is an essential feature of what it is for the Turing machine to represent them. It will turn out that, on the account we think is best, representation could be mental, computational, or any other sort, and is thereby far less demanding than other accounts of mental representation. So, even were there to be no "mental" states in any artificial computer, these computers could very well represent the world in ways that suffice for robust agency.

There is more to be said here if computation can do the work we need it to. Before we say more, it will be instructive to hear from others who are trying to understand the nature of current machines and to determine their status as agents.

Your Tasks

Test Your Understanding
1. What is an algorithm? Can you think of a simple algorithm you use in everyday life?

2. How would you describe a computer to someone who has never used one before?

3. Define the Church–Turing thesis.

Reflect or Discuss

1. There is a difference between having representations and representing. What is that difference, and why does it matter?

2. Can computers be clever or understand? Why or why not? Does it matter? How is cleverness or understanding or sophistication of capacities related to agency?

3. How would you evaluate the quality of our objection to Searle's point that computers merely implement programs? Does the objection persuade you? Why or why not?

Expand Your Thinking

1. Spend some time looking into the history of Ada Lovelace, one of the prominent figures in the history of computing. What might Lovelace have to say about the kinds of machines we have around us today?

2. Some scholars (e.g., Johnson and Miller 2008) have challenged the notion that computation is the right framework for understanding how the world works. Either read the Johnson and Miller article or find a similar philosophical article that defends this kind of thesis. Then, as charitably as possible, interpret their argument and evaluate how persuasive you find it to be.

3. Imagine you had to convince someone that humans are computers. What are some compelling reasons you could offer in support of that thesis? What are some potential objections that your interlocutor might pose?

Further Reading

Boolos, George S., John P. Burgess, and Richard C. Jeffrey. *Computability and Logic*. 5th ed. Cambridge: Cambridge University Press, 2007.

Essinger, James. *Jacquard's Web: How a Hand-Loom Led to the Birth of the Information Age*. Oxford: Oxford University Press, 2007.

Hollings, Christopher, Ursula Martin, and Adrian Rice. *Ada Lovelace, The Making of a Computer Scientist*. Oxford: The Bodleian Library, 2018.

Johnson, Deborah. *Computer Ethics*. Hoboken, NJ: Prentice Hall, 1984.

Johnson, Deborah G and Keith Miller. "Un-making Artificial Moral Agents." *Ethics and Information Technology*: 10 no. 2–3 (2008): 123–133.

Pasquinelli, Matteo. "Three Thousand Years of Algorithmic Ritual: The Emergence of AI from the Computation of Space." *e-flux Journal* 101 (2019). https://www.e -flux.com/journal/101/273221/three-thousand-years-of-algorithmic-rituals-the -emergence-of-ai-from-the-computation-of-space.

7 Agency in Contemporary Machines

Introduction

We are not the only ones trying to think through what agency might entail, generally speaking, or in artificial systems, more specifically speaking. There is, in fact, a burgeoning literature in AI that examines which qualities current machines would need to possess for them to constitute agents or for them to be more or less agentic. Here, we will pause to highlight some recent work by scientists and engineers and scientifically minded philosophers as they grapple with the question of machine agency in real time. The very latest work on agency, specifically directed toward understanding machine agency, is thought provoking. Much of this work is appearing just as this book is being prepared for publication and so has not had time to percolate fully through the various layers of the conversation. Still, we think it is worthwhile to work our way through the main flows of ideas here both to make contact with those whose practice is most connected with possible machine agents and to give a sense for how a minimalist approach can help to comprehend and perhaps unify the diverse viewpoints that now exist.

In addition to finding fascinating many of the various conceptions of agency that are sprouting in the wake of LLMs' rise in popularity, there is another reason to look at these conceptions: we will want to engage more with folks who are skeptical about the possibility of machine agency. One of our important tasks here will be to see whether our own view is compatible with the new AI agency literature or whether, perhaps, that literature has better approaches for sidestepping some of the skeptics' concerns.

Some thinkers are interested in questions of agency because, they believe, agentic AI may be dangerous or a security risk or a system of concern. Put differently, they care whether AI can be agents in the world because they care whether AI will pose a threat to humans and humanity. This is a different starting point from those philosophers who are interested in the question of agency because they are trying to ascertain which entity is responsible for an action, although the two conversations can come together. This difference in orientation is crucial because the purpose for developing a theory—what that theory is *for*—can impact how one evaluates the goodness of that theory. Those thinkers who are primarily concerned with whether AI is a safety risk may be less concerned with sharply dividing AI systems between those who are acting versus merely behaving and may (as some already do) loosely use the same word to capture both kinds of systems. By contrast, thinkers who are concerned with issues of responsibility attribution care deeply about which entity, in fact, acted. We are in the latter category, but we hope our contributions are nevertheless useful to those in the former.

In the discussion that follows, we focus on the conceptualization and definition of agency; we will return to the implications, ethical and otherwise, of adopting one view or another later in the book. Most authors in this literature also believe, as we do, that agency ought to be distinguished from moral agency and that agency is distinct from consciousness and intelligence, respectively. While we will come back to these other concepts, we will not consider them here.

I. Other Views of Machine Agency

To evaluate the agency of contemporary machines, there are many different approaches that are worth looking at. We have in mind, in particular, Carlsmith (2022), Chan et al. (2023), Dung (2024), Lazar (2024), and Shavit et al. (2023). As many differences as these promising views present, their core understandings coalesce around a few main features. We will name these features "goal seeking," "autonomy," and "task complexity," although they go by a number of different names in the papers we have mentioned.

Central to all of the views is goal seeking, which is a term used frequently but left somewhat underspecified and vague in many of the accounts. Carlsmith is clearest about what it means. He calls a system "agentic" "if it makes and executes plans, in pursuit of objectives, on the basis of models of the world" (Carlsmith 2022, 8). How does this work? Carlsmith takes as his paradigm a certain kind of human cognition, which has two crucial features:

"(a) using a model of the world that represents causal relationships between actions/outputs/policies and outcomes to (b) select actions/outputs/policies that lead to outcomes that rate highly according to various (possibly quite complicated) criteria (e.g., objectives)" (Carlsmith 2022, 9). He goes on to suggest that there are algorithms with this capability that "implement explicit procedures in the vein of (a) and (b)—algorithms, for example, that search over and evaluate possible sequences of actions, or that backwards-chain from a desired end-state" (Carlsmith 2022, 9).

Goal seeking, then, for Carlsmith can involve planning, and that might seem to go beyond the conceptual resources of the minimalist view. What planning on the basis of a model of the world amounts to, however, is deploying items in a behavioral repertoire under guidance by various representations, one of which is the representation of the relevant states of the world, and others of which are of various intermediate states of the plan. In that sense, Carlsmith's view is, we think, on the right track. It is worth noting, though, that he does not count some things as agents that we think ought to be. He writes, "MuZero—a system which learns a model of a game (Chess, Go, Shogi, Atari) in order to plan future actions—qualifies as an agentic planner in this sense, as do AlphaZero, AlphaGo, and (at least on my current understanding) various self-driving cars. Thermostats, bottlecaps, forest fires, balls rolling down hills, and robots twitching randomly don't qualify—they're not doing something close enough to planning, using a model of the world, in pursuit of the outcomes they cause" (Carlsmith 2022, 10). As you know, we think thermostats count as agents, but despite what Carlsmith *says*, we also think that, on his own view, they should count as agents because their representation of the temperature is a model of (part of) the world, and its setting counts as an objective. No system models the world completely. The important thing is to be modeling the parts that are relevant to one's purpose, and the thermostat does that.

Autonomy is also an important component of most of these views. We agree that it has a crucial role to play in whether we regard some system as an agent. For example, if you see a drone in the sky, you cannot know whether it's an agent until you know whether it's being remotely controlled. Many of these theorists are similarly worried about whether humans are "in the loop" of some activity and what role that plays in whether and how effectively the activity gets done. That is an important consideration. Some of these authors go too far and include an explicit evaluation of the efficacy of various systems into their evaluation of whether they can count as agents. Shavit et al. are concerned with what they call "independent execution." By this, they mean "To what extent can the system reliably achieve its goals with limited human intervention or supervision?—Example: Cars capable of level 3 autonomous driving, which can operate without human intervention under certain circumstances, have greater independent execution than traditional cars that require continuous human operation" (Shavit et al. 2023, 4). Chan et al., meanwhile, call this "directness of impact": "the degree to which the algorithmic system's actions affect the world without mediation or intervention by a human, i.e. without a human in the loop" (Chan et al. 2023, 4). Dung, who explicitly separates autonomy and efficacy, writes of this feature, "A being is efficacious if its goal-directed behavior affects the world, without someone else's mediation or intervention. Efficacy encapsulates the idea that a being is more agentic if its behavior does not require support by others to matter . . . The efficacy of a system can be specified more precisely by the extent to which it requires another being, typically a human, to be in the control loop" (Dung 2024, 5).

To be sure, it would be strange to consider something an agent if it lacked the behavioral repertoire to work toward its goals or ever to achieve them. But adding in the requirement that there be little to no outside involvement of other agents in their activity diverts us from questions about agency to contingent matters of causality. And that sounds wrong to us.

Think, for example, of one of the most powerful people in the world as this book goes to press. No, not Elon Musk. Rather, we're thinking of Taylor Swift. And here's the thing about Taylor Swift: she is astonishingly effective. When she comes to town, the economy of the town is significantly and positively impacted. Her actions have profound downstream impacts on the lives of many, many people. But her ability to do any of this *without the assistance of others* is quite limited. Her voice requires amplification at her concerts, her crew makes the venues appropriate for these occasions, streaming services and radio bring her voice into her fans' ears. In terms of *direct* causal impact, Taylor Swift is not much more effective than either of us. But when she engages the power of other agents in her surroundings, her impact is huge. (The same, by the way, is true of Elon Musk.)

Many of these authors are also focused on task complexity. While the capacity to succeed in very complex endeavors is a good measure of the intelligence, sophistication, or intricacy of any dynamical system, it is left unclear why it should be a measure of agency itself.

Certainly, agents who can perform a wider range of complex tasks are thereby more interesting and worth understanding, but their agency—the difference between them and non-self-directed systems—seems not well captured by the complexity of tasks *as such*. But that is not the key reason to be suspicious of the complexity measure as used by these authors. Rather, that comes from a general problem of orientation with these views overall.

While these various authors have captured some core features that are connected to agency, there are two important things to note: first, human-centrism is built into many of the approaches (subtle though it may be), and second, they structure the discussion in a manner that lets judgments about which systems are agentic drive their analyses of agency.

II. Worries about These Views

Let's consider first the problem of human-centrism. OpenAI, the company that brought us the GPT series and that promises to make AGI for the good of humanity, does a lot more than just engineering. They also, like other AI labs, have policy teams dedicated to understanding various consequences of AI for the world. One team working on understanding how to control AI begins by trying to understand what has come to be called "agenticness." The team defines it as "the degree to which a system can adaptably achieve complex goals in complex environments with limited direct supervision" (Shavit et al. 2023, 4). This strikes us as a really good proxy for agency, in many circumstances, especially given the kinds of machines we already have. It is difficult to see how a non-agent could succeed at this.

A familiar worry arises. It was also very *difficult to see*, in the past, any nonhumans as agents, and that colored past analyses of agency in unfortunate ways. That we find something *difficult to see* is unlikely to be helpful in understanding what is going on in novel domains. What is needed, again, is an account that does not rely on our antecedent sense of agency, and that is not human-centric. Of course, beginning with human capacities is not a bad idea, as that gives an initial sense for the salient features of agency. However, framing agenticness around human capacities, as the OpenAI team does, is a less promising route. They characterize an important feature of agency, goal complexity, this way: "How challenging would the AI system's goal be for a human to achieve and how wide of a range of goals could the system achieve? Properties of the goal may include target levels of reliability, speed, and safety" (Shavit et al. 2023, 4). But this is not the right way to fold complexity into agency. Special-purpose machines can be devised to achieve almost any goal that a human can achieve and can do so without any agency at all. That is what we learned from earlier iterations of chess engines. These machines reliably and with little difficulty defeated the very best chess players in the world. But they were pure computational engines without anything that could be recognized as agency. As we noted earlier, they simply were not playing chess, although they achieved its end state easily.

OpenAI is not unique in offering a human-centric approach. Dung proposes a five-dimensional account of agency that is meant to allow for a nuanced characterization of how agential a given system is. One thing that stands out sharply is Dung's fifth criterion, *intentionality*. Intentionality, recall, is the way our thoughts get to be about the world. It will be important for any conception of machine agency to show how the representations guiding machine behaviors can themselves be appropriately about the world. So, Dung is making the right move. The problem is that he pivots back to self-reflection and a kind of felt engagement as characteristic of intentionality. He is focused on "the possession of beliefs and the ability to reflect on one's reasons as such and to revise or endorse them on this basis" (Dung 2024, 14). But that goes beyond aboutness and instead centers *human* features, the way our own intentionality works, in the general notion of agency.

The second problem with the accounts is that they are largely focused on exemplars for determining criteria of agency. That is, their authors have some examples of agents in mind and are attempting to pick out from them those features that are most associated with those examples. As we saw earlier in the book, when we made the first attempt to build out a "little theory" of agency, that is often not a bad way to proceed. Our intuitions in many domains are quite powerful and reliable, and we are pretty good at using them to group things together and pick out their similarities. You might call this way of proceeding "diagnosis as conceptual analysis," seeing agency in the world and attempting to find its source. The main problem with this approach in the current context is that we do not have many clear, uncontested examples of machine agency to diagnose, and we seem to be outside of the realm of agents where we know our intuitions are reliable. That is, in part, why we opted early in the book to try to understand agency independently of any intuition.

So, what does our own account say in this context? We have talked a lot about the thermostat. But how does the minimalist account do when confronted with a more fully realized computationally powerful candidate for agency? Let's consider one. Alpha-X is a chess-playing engine unlike those that came before. Earlier types of engines were very strong on tactics—that is, they are very good at evaluating the present state of the game and determining whether sequences of moves exist that end up with a new position whose state appears advantageous from the point of view of having more and more valuable pieces on the board. The move with the highest score is selected. That's a simplification, but chess players at very high levels have known for a long time how to defeat most such engines by thinking about long-term advantages and the more dynamical features of the game. Of course, even so, a powerful enough calculation device will eventually, by sheer power, defeat even the most creative human player, as we saw starting with Deep Blue. But it will do this by brute computational power, not by deploying various behaviors in light of representations of the world and its goals.

Alpha-X is different, and the clearest way to express this difference is to say that it behaves in a way that tends toward long-term advantages rather than short-term material superiority: it behaves *strategically*, deploying various tactics in service to its long-range goals, bringing its representation of the state of the game in line with one where it is winning. The story of most chess engines' beginnings is that they are loaded with a great deal of factual information so that, for the first few moves, they are programmed with the best move to reply to whatever the opponent's moves are. Then, they are given a great deal of computing power and a kind of algorithm for weighting the goodness of various sequences of moves along with rules for which sequences to consider. No human can keep up with the calculations, but we can reasonably understand what the calculations are and what is at issue at each moment. In contrast to these engines, Alpha-X was simply programmed with the rules of chess, the capacity to update its playing algorithms, and was set to play many, many games against itself. What emerged from that was a strategic capacity that chess theorists find remarkable and beautiful. Moreover, their experience is that, for the first time, a computer was "playing chess" rather than executing some program that made it impossible for humans playing chess to defeat it. That's all pretty metaphorical, but the main point is clear: Alpha-X is strategic, not merely tactical, and, on our view, is an agent.

Another kind of system to test our theory against is LLMs. At least the others analyzing machine agency are quite taken by its prospects. We began work on this book before ChatGPT came out, but an important prompt for us was its predecessor. We were struck by the apparently sophisticated reasoning of GPT-3, as well as its capacity to harm. Once we had settled on our view of agency, we reevaluated GPT-3 against the view and saw that, in fact, it has no agency. GPT-3 seemingly understands a lot of things, but it does not really; it simply grinds up massive amounts of data and spits out the result of a statistical weighting. It makes no use of representations of the world's states, its states, or anything else. Here, we might say that our way of proceeding is "conceptual analysis as diagnosis"; we begin with the concept and then assess existing systems against that.

Looking holistically at the current conversations on agentic AI, we, by and large, find these to be helpful. The main virtues are that they have identified many features that we can look to for evaluating the agency of dynamical systems we might run across in many different contexts, and they have appropriately (in our opinion) centered agency in the discussion of the emerging capacities of artificial systems. We think their discussions can benefit from a concept-centered approach to agency of the sort we have offered, and that, we hope, will refine the intuitive understandings currently at play.

III. Unseeing Agency in LLMs

Not everyone will think that any computational model of agency—either ours or the ones advanced by others—does the work it needs to do, and we should take their objections seriously. While we cannot consider all objections, we do want to discuss a very important one, called the "Octopus test." It provides a powerful argument against seeing agency in LLMs. On the surface, the Octopus test seems like another iteration of Searle's so-called Chinese Room argument. Even its authors seem to think that, but it goes beyond it in important ways.

At the end of the last chapter, we rejected Searle's conclusion, but our rebuttal of Searle against computationalism only showed that his *argument* was not good. Moreover, Searle's influence remains powerful in the discussion on AI, especially among those who think carefully about the difference between ersatz and genuine language use. Many folks think that the latter is essential for robust agency, and that is something we do not dispute. But what's really going on here is that the authors we have been considering in this chapter, among others, have "seen" language use in LLMs. From there, they have seen the road to artificial general intelligence and robust agency. But unless LLMs are *really* using language, that road is a dead end. LLMs are astonishingly good at taking in inputs and generating outputs that seem just like conversation. But do they really use language?

Not in Searle's sense, you might think. Why not? This is because LLMs still do not have intentional states that would allow their words to be about the world. LLMs merely transform strings of syntax into other strings of syntax using very complicated statistical algorithms applied to vast quantities of speech on which they have been trained. But perhaps that's not quite right. After all, unlike the person in the so-called Chinese Room, LLMs do not work by simply looking up responses in a codebook. Instead, they actively generate their responses. So, why is it wrong to say that they are really using language?

Two computational linguists, Emily Bender and Alexander Koller, have explained the problem in a dramatic fashion. That's where the new thought experiment, the Octopus test, comes in. Here's how it works. Two English speakers, A and B, are stranded on separate desert islands. There are telegraph machines on these islands that happen to be connected together but not with any other islands or the mainland. But that's okay; A and B seem happy enough to be able to communicate with each other. What they do not know is that the cable that connects the islands under the ocean is being monitored by O, an octopus, who does not know anything about English. Over time, O gets pretty good at anticipating what B will reply to any of A's remarks.

After some more time, O decides to cut the cable and pretend to be B. If O is as good as the iterations of various LLMs that already exist as we write, then O can probably do a good job fooling A and convincing her that O is actually B. So far, so good. What O cannot do is understand what is being said, and so when a completely new situation arises, when A invents something new and describes it to B and asks for help improving it, O is helpless and can only offer platitudes of congratulation. And when A, for example, is in trouble and needs some help resolving an unexpected emergency, O also has nothing to contribute. This is all because O only makes statistical predictions of what B would say, and without content that is relevant to the situation, there is nothing to predict (Bender and Koller 2020, 5188–5189).

One important distinction, then, between O, the octopus, and a real language user, like you, is that your use of language comes with what are sometimes called "language entry and exit rules." For you, language is a tool that, in part, allows you to express facts about the world (including your own mental states) to others and to learn similar things from them. But it is not your only tool. Instead, it works in concert with your own experience of the world itself. There is steady feedback between you, the world, and your language use that ChatGPT does not have. Now, that all seems right. So, it would be premature to attribute genuine language use to any extant version of GPT or any LLM for that matter.

Some of the people who are working on the dreams of AI agents, even as this book is being finished, are specifically addressing the problem of language entry and exit and training AI on worldly systems. What will happen when these AI are expanded to include that worldly input is likely to be quite exciting and may well change our understanding of the scope and limits of AI.

Still, Searle, at least, would not believe that incorporating worldly feedback or any other learning scheme would suffice to allow for AI to understand language, no matter how tightly connected to the world they are. But it is less clear whether Bender and Koller are committed to such a strong conclusion. While the Octopus thought experiment is presented as an extension of Searle's argument, it is much narrower in scope; it focuses more pointedly on the difficulty of getting computers to have the intentionality necessary to have their computations be about the world. Bender and Koller's argument relies crucially on the computers involved learning only from predicting the language use of other systems. But if they become trained on predicting the world itself, or representations of it, then what?

More is needed to answer this question. Even if Searle is wrong, even if there is more to a computer program's implementation than he thought, and even if we could make machines with all the sensitivity and activity we want, there could still be the problem that those machines are not in touch with the world *in the right way*, and what they are "talking about" is not really the world but simply the play of symbols that pass through them. Remember, while we argued that Searle is wrong about the nature of implementation and that there is something more going on in the case of computer programs than mere syntax, we did not back that up with any analysis of exactly what that more is. We will need more tools to do this analysis, and we think we can find them in communication theory.

IV. Looking Forward

We have put computation at the center of our discussion, and some do use a computational model alone as a theory of agency. But computation, as the above worries have made even more clear, will have to be only a part of seeing how the minimalist account of agency gets implemented in machines. That way of seeing things is, so to speak, from the inside out. Computing by itself is important, but it is not enough. The fact that some system computes does not by itself make it appropriate to think of it as an agent. As we saw, agency requires linking representations to behaviors in the right way. While making those links is computation, even in the case of the thermostat, the structure that connects the agent to the world is more than the computation itself. It is interesting to note that even if Alpha-X is an agent, its impact on the world is not even, at the end of the day, as profound as is the thermostat's. All of that computational sophistication did not really make it work better *as an agent*. The computational advances it incorporates do not do much toward making its states more worldly directed.

There seem two ways to go right now: explore more deeply what could be wrong with any view that ignores the significance of our own (and any other agents' we know) biology, or fill in the gap in the story of computation and machine agency to include tools needed for intentionality, the capacity for their states (and linguistic utterances) to be about the world in the right way. We will do both, starting in the next chapter with biology and following that with a discussion of communication theory.

Your Tasks

Test Your Understanding
1. What is the difference between "autonomy" and "agency"?
2. Define "goal directedness" and, in your own words, explain how it is connected to representations.
3. What happens in the Octopus test, and what are the implications of the thought experiment?

Reflect or Discuss
1. Do you think Alpha-X is an agent? Why or why not? What additional information would you need to make this assessment?

2. Why might "diagnosis as conceptual analysis" not be the best approach for assessing agency? What are some ways to defend the approach?

3. Is it possible to convince Searle that LLMs, if changed in the right way, could have intentionality? Assuming we could improve our existing LLMs in some way, would we be able to persuade Searle?

Expand Your Thinking

1. Choose an AI system and get prepared to do a little bit of research. Here are some we have not covered in the book that you might be interested in and could use for this activity: GATO, CICERO, YOLO. Once you have chosen your system, write a summary of what the system can do, what it is for, and the kinds of activities it can engage in. Then, assess it on goal seeking, autonomy, and task complexity (remember to define your terms). Finally, look at the notes you have taken on the system and consider, more holistically, whether you think the system is an agent. If you are struggling to answer the question, here are some steps that can help you with your analysis. First, make a list of the features you think an entity needs to have in order to count as an agent. Each feature should be individually necessary, and all the features taken together should be sufficient for the constitution of agency. Second, check whether the AI system has all of the features you've listed. Third, see what surprises you, seems intuitively plausible, strikes you as wrong, and so on. Pay attention to your reactions, and ask yourself whether you have good reasons to have the reactions that you do.

2. Irrespective of whether LLMs are agents, they may still be able to cause harms of various sorts. A seminal paper in the study of LLM harms is "On the Dangers of Stochastic Parrots: Can Language Models Be Too Big?" by Bender et al. Read this paper and then assess how the LLMs you see around you are likely to cause ethical harms.

3. Some thinkers worry about what is called "existential risks." These are risks that pose an existential threat to humanity. Within this group, there are some who believe advanced AI may pose such a risk. Read the Carlsmith paper we have introduced in this chapter and consider the strength and weaknesses of the view that certain kinds of AI (including agentic AI, which is Carlsmith's target of analysis) can pose an existential threat.

Further Reading

Anderljung, Markus, Joslyn Barnhart, Anton Korinek, Jade Leung, Cullen O'Keefe, and Jess Whittlestone. "Frontier Ai Regulation: Managing Emerging Risks to Public Safety." *OpenAI*, July 6, 2023. https://openai.com/research/frontier-ai-regulation.

Bales, Adam, William D'Alessandro, and Cameron Domenico Kirk-Giannini. "Artificial Intelligence: Arguments for Catastrophic Risk." *Philosophy Compass* 19, no. 2 (2024): e12964.

Bender, Emily, and Alexander Koller. "Climbing towards NLU: On Meaning, Form, and Understanding in the Age of Data." In *Proceedings of the 58th Annual Meeting of the Association for Computational Linguistics*, 5185–5198. Association for Computational Linguistics, 2020.

Carlsmith, Joseph. "Is Power-Seeking AI an Existential Risk?" (2022). https://arxiv.org/pdf/2206.13353v1.pdf.

Chan, Alan, Rebecca Salganik, Alva Markelius, Pang, Nitarshan Rajkumar, Dmitrii Krasheninnikov, Lauro Langosco, Zhonghao He, Yawen Duan, Micah Carroll, Michelle Lin, Alex Mayhew, Katherine Collins, Maryam Molamohammadi, John Burden, Wanru Zhao, Shalaleh Rismani, Konstantinos Voudouris, Umang Bhatt, Adrian Weller, David Krueger, and Tegan Maharaj. "Harms from Increasingly Agentic Algorithmic Systems." *FAccT '23: Proceedings of the 2023 ACM Conference on Fairness, Accountability, and Transparency* (2023): 651–666.

Dung, Leonard. "Understanding Artificial Agency." *The Philosophical Quarterly* (2024). DOI: 10.1093/pq/pqae010.

Lazar, Seth. "Frontier AI Ethics." *Aeon*, February 13, 2024. https://aeon.co/essays/can-philosophy-help-us-get-a-grip-on-the-consequences-of-ai.

Shavit, Yonadav, Sandhini Agarwal, Miles Brundage, Steven Adler, Cullen O'Keefe, Rosie Campbell, Teddy Lee, Pamela Mishkin, Tyna Eloundou, Alan Hickey, Katarina Slama, Lama Ahmad, Paul McMillan, Alex Beutel, Alexandre Passos, and David G. Robinson. "Practices for Governing Agentic AI Systems." *OpenAI*, December 14, 2023. https://cdn.openai.com/papers/practices-for-governing-agentic-ai-systems.pdf.

8   Biology and Brains

Introduction

At this point, you might be worried that we are not being fair to folks who think that substrates are important when thinking about agency. That is probably not unreasonable. We have asked you at various points to think about what you regard as important in an account of agency. Many of you, reading the last few chapters, will have been pretty sure that such concepts as "mind" and "consciousness" and "feeling" are very important parts of the concept of agency. Our students over the years have certainly thought them important, and they *are* important. At least, they're important to what agency is in humans, and that is the best example we all know about when trying to understand it.

Earlier, when we discussed Searle's idea that the brain that makes the mind must have certain causal powers, we suggested that he was ultimately demanding some kind of *intentionality substance*. We think that looking for things like that is not fruitful, but we are open to the idea that there may well be more to it than that. So, maybe when Searle demands that there be certain kinds of causal powers in the materials that make up brains that are capable of producing minds, the mistake is wanting the powers to be in the *materials* rather than in the *organization* of the materials. It makes sense to a lot of people to think that the organization of various materials is doing more work than the particular materials themselves. It might also seem like if organization is all that matters, then there is no objection to the idea of creating machines that do think, act, and have intentionality after all.

On the other hand, maybe there is more to organization than meets the eye and more to causal powers than we might have thought. This chapter looks at the substrate question from a different direction. Maybe there is something about the particular evolutionary history of biological systems, and the way they are embedded in their biological context, that is required for mind and consciousness and thus, if agency generally is like agency for people, for agency itself.

I. Agents as Organic Unities

What if minds *are* necessary for agency, and *biological* organization is necessary for minds? Here, we will see one example of how philosophers using biology and complexity theory have tried to understand the nature of human agency still through the lens of human minds. For these thinkers, minds are best thought of not as separate from the body but instead as features of certain kinds of complex dynamical systems: the biological.

Here's how to think about complex dynamical systems. Start with simple systems. Some systems, even some with very simple basic laws of operation, have behaviors that are impossible to predict fully. While the laws are simple, the systems are so structured that even the smallest changes in starting conditions lead to very large changes in their outcomes. Think about trying to balance a sharp pencil on its tip. That's pretty much impossible, and it will definitely fall over. What is also impossible is to observe exactly how it is standing when you let go so that you can predict how it will land, say whether the eraser will be pointing to the right or to the left. Of course, you can set it up to make sure it goes one way or another. But setting it up as straight as possible and then letting it go with very little disturbance will make it seem random which way it goes. But it is not random; it is simply a system whose behavior is so sensitive to its starting conditions that no changes to them, however small, are irrelevant.

Things get even more interesting when there are lots of bits interacting together to produce the behavior. In addition to the kind of unpredictability of the pencil (what is called its "chaotic" behavior), there are interesting new things that show up—things that do not *seem* to be reflected in the equations of motions of all those interacting bits. Studying all of that is the business of dynamical systems theory.

Philosopher Alicia Juarrero uses dynamical systems theory to give an account of action and intentionality. In *Dynamics in Action: Intentional Behavior as a Complex System*, Juarrero argues that agents are best conceptualized as complex adaptive systems that exist in a particular biological context. Juarrero sees dynamical systems theory as the key to account for all the many interesting things that happen when, for example, your entire body is set in motion by a very tiny change in the configuration of one neuron that cascades through the entire complex dynamical system that you are.

But do we really need to understand dynamical systems in order to understand people acting in the world? It seems like we can understand people, at least, by thinking about belief–desire pairs as controlling behaviors. Juarrero disagrees. Conventional theories of action of that sort, Juarrero thinks, are wedded to a limited conception of cause understood as mechanistic, efficient cause. On Juarrero's view, modern philosophy took the worst from Aristotle. We embraced the erroneous view that there is no such thing as a self-cause—that is, Aristotle, and subsequently the remainder of the Western tradition, holds that nothing can move, cause, or act on itself in the same respect that it is being acted on by itself, since it would have to be both active and passive at the same time. But while Aristotle's story of causation had a complex structure with multiple kinds of cause, current theories of causation have only one. Aristotle gave us formal, material, final, and efficient causes. (Here's an analogy to help differentiate the causes. Formal cause is the plan, like an architect's drawing of a house. Material cause is the stuff, like the wood the house is made of. Final cause is the reason, like to have a place to live. Efficient cause is the mechanism, like how you actually build the house.) Nowadays, in most discussions of causation, we have only the efficient.

If all causes are efficient causes and we think of all interactions basically like the banging together of billiard balls, then, Juarrero says, you have the standard way folks think about human behavior. Intentions, beliefs, and desires cause bodies to move in the same manner that one pool ball causes another to move. It is all very simple and all very predictable. But she thinks this is a weak notion of explanation and completely inadequate to understand self-cause.

On Juarrero's own view, standard accounts of action theory that are anchored in these misconceptualizations of cause and explanation inevitably fall apart because they are atemporal and acontexual. Moreover, they ignore the role of self-cause and thereby leave a "gap" between cause and effect, rendering the story incoherent. We are not entirely sure that Juarrero gets the history exactly right, nor that she gets exactly right how most are thinking about action explanation. But she is right that self-cause, what some philosophers call "self-movement," is an interesting and difficult subject.

The answer to our question above, about whether we need to understand complex dynamical systems in order to explain human behavior, may well be "no." However, unless we pay attention to the insights from dynamical systems theory, we are unlikely to understand how it is possible that systems that *are* apt for belief–desire pair explanations could possibly come about. And we need to understand *that* to evaluate the prospects for generating an appropriate analogue of human behavior. We also just think that Juarrero's method of bringing dynamical systems theory to bear on the problem is a philosophical view worth explaining. Here is how it works.

II. Complex Adaptive Systems

What is a complex adaptive system, and what makes it special? In addition to the sensitivity to initial conditions, and the complexity that comes from having many interacting subparts that we saw in complex systems, complex adaptive systems have something else. They have what is known as *emergent* behavior. How come birds flock together? One obvious thing to think is that there is something that all the birds want to do, like fly around as one organized group. But there does not seem to be any such mechanism in operation when birds flock, no sensitivity that each bird has to know how well they are all flocking together, and no behavior that is directed at correcting any failure to flock together. Instead, it appears that each bird is sensitive to what the several birds closest to it are doing, and it adjusts its flight to try to maintain a given distance between itself and the others, or something like that. The exact thing they do seems to be a little unclear in the literature, but what is clear is that they are only paying attention to their near neighbors, not the group of birds as a whole. This very local, causal interaction between each bird and its near neighbors in the group somehow spontaneously results in flocking, where huge numbers of birds are together flying in a single group with what looks like a definite, shared purpose. If you have not seen this yourself, we recommend that you look up some videos on bird murmuration. It's quite amazing. Eerie patterns and long-range correlations appear and then disappear.

Once you start looking for other examples of complex adaptive systems, they turn out to be all around. For example, the group behaviors of social insect colonies reveal that these colonies are complex adaptive systems. In a bee colony, no individual bee, worker or queen or drone, has a vision for how the colony will build a hive, when to attack a potential threat, or when to swarm and generate a new subcolony. These behaviors emerge spontaneously from the small-scale interactions between each member of the colony and the pressures the whole colony faces from the larger system it is embedded in. Flocking birds, social insect colonies, and even global supply chains exhibit these kinds of spontaneous behaviors arising out of their complexity and adaptation to the larger environment. Then, there is the case of humans. We are complex adaptive systems, and our brains are as well, and the entire organism that each of us is exhibits quite remarkable emergent behavior.

Juarrero uses her framework to illustrate how humans, conceptualized as organic unities embedded within a particular time and context, can act intentionally. For Juarrero, the only way to make sense of the notion of acting intentionally is by offering an account of how it is that organic unities like us can have control over their own behavior, which just is acting intentionally for her. The agent, in this framework, is the entire complex adaptive system rather than a disembodied mind attached to a body. Even so, the framework allows for minds with causal powers as emergent features of the system. The dynamics of such systems "provide the framework for the behavioral characteristics and activities of the past" (Juarrero 1999, 106). Human beings, now seen through the lens of complex adaptive systems theory, exhibit characteristic self-organization, akin to the flocking of birds, but expressed as coordinated bodily behaviors directed toward some end. Like any other organized wholes, human organisms gain powers that are not present in the simple structures that they are constituted by (individual cells). While every action is causally determined, somewhere deep in the physics, it is impossible for external observers to determine what future action will be taken. Of course, many of our actions are predictable in a general sense, but detailed billiard ball–style analyses are simply unavailable.

Dynamical systems theory is worth delving into much more deeply, but we do not have the space here. We will focus here on the main takeaway for the question of whether it makes sense to see *mind* as the key difference between systems that can act intentionally (like us) and other systems that cannot. Juarrero is telling us that instead of extra properties or capacities that exist within or alongside some entity that make it *be* an agent, it is the complicated connections between all the parts of the entity that allow it to *express* agency.

For Juarrero, before we can unpack what intentionality is, we must better understand how agents, as organic unities, actually function. The self-organizing system exhibits inter-level causality, both bottom-up and top-down. Juarrero tells us that this variety of levels of causation allows us to see organisms as both agent and patient, as the source of the doing and the target of what is being done, all at the same time. This notion of "self-cause" draws on Aristotle's concepts of formal and final cause to argue that the causal mechanisms at play within the organic unity can best be understood within the operations of constraints—that is, organisms are both structured according to a *plan* that is implemented, interactively, within their environmental and internal biochemical contexts, as well as being goal directed, where navigating toward those goals happens within environmental and biochemical contexts. These contexts constrain the possibilities of the organism, making some behaviors impossible and some goals unattainable. Humans can only grow so large, run so fast, and carry so much.

But in addition to limiting the organism, these constraints open up degrees of freedom. As an organism has more and more complex subsystems, it gains capacities to move within a space of constraints that allow for self-governed action. For example, to the extent that we humans have some intention to do something, we generate through that intention a top-down constraint that reshapes the dynamics operating on the lower levels. To be a cell is to be bounded in space, but only through that boundedness can there even be such a thing as a one-celled organism. And such organisms, over long evolutionary time, are bound into multicellular organisms. This limits individual cell autonomy but vastly opens up the range of activity of the new multicellular systems. In specific multicellular organisms, like us, developmental programs in our cells constrain development to specific biological pathways. This both limits the autonomy of the blastula and makes possible its transformation into a human. In addition to the internal structure of the cells themselves, we are finding more and more that the specific biochemical context of the gestational environment constrains the possibilities of development at the same time it makes development possible.

As constraints limit the number of ways in which the components of the system can be arranged, altering what is possible and what is not, they also change the probabilities of the options that remain possible. Think about this in the context of flocking birds. If the birds simply fly about wherever the mood strikes each individual bird, the probability of all the birds flying to the left at the same time is very, very low. But if the birds are looking to their close neighbors to decide where to turn, that probability is much greater. The way we would put this in information theory, which we will explore in the next chapter, is that these constraints reduce randomness "by altering the equiprobable distribution of signals, thereby enabling potential information to become actual information" (Juarrero 1999, 82). As with overt behavior, so also with thought and language. Sense and meaningfulness are only possible within a system bounded by constraints. Constraints also enhance freedom and indeed make it possible. This is true for language, activities such as jazz and ballet, and, crucially, for organisms like us. If we conceptualized constraints only as restrictions, then the more complex an organism becomes, the less it would do. A blastula, having fewer constraints, would have more degrees of freedom. But this notion is false. By constraining the blastula to become a human, these constraints allow for an exercise of capacities that could never be exemplified by a random collection of cells.

For most of the time, the complex system operates within the space of the internal constraints. Constraints directly impact action. For self-organizing systems, Juarrero suggests, the way that they organize gives them determinate sets of interests. What manifests itself as an option or choice will depend on the complex system's history of interactions. Consider the following. Suppose someone's developmental history takes place in a context in which crickets are meaningfully read as food. Then, eating crickets as an afternoon snack is an option that will register for her. For others who do not conceptualize crickets as food and who visit Thailand and see this as a food choice, there might be a different reaction. Our behaviors, as with all complex adaptive systems, are conditioned by networks of past contingencies and the structure of the world within which we are embedded.

Juarrero suggests that if we see our developmental context as a kind of landscape of options, much of our behavior can be understood as following the path of least resistance on some given developmental trajectory. At the very top of a hill, for example, there are many ways to head down. But the hill is not completely flat from side to side, and as one heads down the hill and the paths become farther and farther apart, it becomes more and more difficult to move from one path to another. The ridges that grow up along the paths and keep those paths separate from each other also keep us from deviating from our own course: they serve to attract us to the pathway we are already on. Much of the time, our bodies simply do what they do, and we act by habit, rolling along down the hill on the path we started on. Indeed, our identities can become calcified over time because the strategies with which the brain processes information and the habituated tendencies we have developed to engage with particular kinds of phenomena become solidified. We often only notice that we are making choices, even that there are choices to be made, when those habits are interrupted. These stable dynamical pathways yield stable modes of behavior of the organism over time. But both incremental and radical change to our pathways of activity is possible.

Consider again the example of someone newly seeing crickets being eaten for food and deciding how to respond. On Juarrero's model, a number of competing neural attractors may be pulling her in different directions. On the one hand, she may be curious to taste the cricket herself; on the other, she may feel a twinge of disgust at the thought. If a sufficient degree of disorder and chaos were in the system (her) as she processed this information, this moment would represent what Juarrero calls a bifurcation. Bifurcation points force us into new spaces of possibility. This is the case when the path branches, and the system (person) must go down one way or another. But there is no clearly determined way to go. Much like the pencil balanced on its point, the situation is unstable, with the attraction of each path seemingly equal. Once enough chaos accumulates, the system (again, our adventurous/timid eater) reorganizes and stabilizes around one outcome, and a choice is made.

The outcome of this bifurcation process is what Juarrero calls a "phase shift": the entity has formed an *intention* to move into a new space, resulting in the radical transformation of the internal configuration of the system, including constraints and the system of attractors. In a phase shift, instead of "rolling down the hill" on the existing developmental trajectory, the entity will "jump" over into another trajectory entirely (Juarrero 1999, 234). Notably, while phase shifts can be externally triggered, they are fundamentally internal. Given this, the existing internal dynamics of a system and the extent to which that system is robust will determine what is likely to trigger a phase shift. For a sheltered Westerner traveling to Thailand for the first time, perhaps seeing people eat crickets is sufficient for the phase shift to occur. For a seasoned traveler, the same experience would fit within the trajectory of "rolling down the hill" and not represent a bifurcation point. Understanding human action thus requires us to look at the dynamics of interacting systems within a biological context (Juarrero 1999, 157). The boundaries between the internal and the external become less significant. On Juarrero's view, the loops that run through and the dynamics that cross borders between brain, body, and world regulate an entity's ongoing behavior.

While there are many places to disagree with Juarrero's account, we see the value of her project and in trying to see how various parts of human agency are to be identified with and grow out of the features of those complex systems. Understanding the way various dynamical systems implement and manage their self-control functions and their responsiveness to the outside world will be key both to understanding whether it makes sense to see them as agents as well as to understand the various kinds of acts they can perform.

III. How Necessary Is the Biological, Evolutionary Context?

Juarrero's view of agency might seem to get us closer to understanding how machine agency might work. If to be an agent is to be a type of complex system, then it's not too difficult to imagine that a machine could be an agent. But Juarrero's account cannot explain machine agency in large part because Juarrero is committed to the view that these entities need to be organic structures embedded in a specific biological context. This is why Juarrero has a lot to say about blastula, simple though they are, and less to say about possible machine agents.

Our minds and the minds of other sophisticated natural agents (e.g., cephalopods) are the byproduct of certain classes of information processing devices. In our experience of the world, we have only encountered one class of such devices that generates minds: brains. As far as we can tell, nothing else in the world makes minds, but it is plausible, at least *possible*, that there are other ways of doing it.

Principally, one thinks of the representational capacities of humans, for example, as a crucial part of what makes us minded. In order to carry out a variety of tasks, we use our brains to process information, and then we use the output of that processing to guide our behavior. Unlike other simpler systems, however, our information processing comes along with an experience of the world, with a feeling for what it is like out there.

It is not clear how minds result from the various stages of processing the information we receive, but it does seem clear that there is nothing special about the materials out of which our brains are made. There are arguments that these materials' mechanical, chemical, electrical properties and so on are important for how they play crucial causal roles in generating our mindedness, but the *particular* materials used seem much less important. (We note here that Juarrero herself thinks that information theory is inadequate to the task of explaining the behaviors of dynamical systems. We will address her objections in the next chapter.) The materials must be apt for the kind of processing we engage in, of course, but beyond that, what does the heavy lifting in making minds is the organizational structure of the system, the brain, itself. That particular kind of system, a mammal or cephalopod brain, is the product of a long chain of evolutionary changes. That evolutionary history encodes a kind of solution to the problem of information processing for the sake of getting organisms to produce offspring-producing offspring. It is unlikely that we can replicate in silicon (or any other medium) such a thing, given our present and near-term technical capabilities. However, the key feature of that solution is its highly structured and causally responsive information-processing architecture. It may be possible to replicate the *sophistication* and the *aptness* for controlling behavior via beliefs and desires that evolved brain architectures have but in novel materials. It may also be that any such replication will by the nature of things produce a mind. But, then again, it may be that such a replication does *not* produce a mind. So we think it is best to reserve judgment about the need for minds as we are assessing the possible agential capacities of machines.

Our own way of being in the world, our thoughts, our hopes, our dreams, our fears are strongly conditioned by our evolutionary history. Our very brains, and the way they go about producing representations of the world (beliefs, images, feelings) are a product of that evolutionary history. One thing about evolutionary history is that it is like any other process of accretion. Things stick around long after their initial purpose has been superseded, and sometimes it is not clear what those things are now for, and even when it is clear, they are often now for very different things than they were originally. Here is one simple example. We have specialized brain cells that are built to react incredibly fast to the presence of snake-like visual impressions. This happens much faster than conscious thought, and it is hardwired in. Why? Presumably because snake avoidance was a very important survival trait a long time ago. This is our evolutionary history continuing to manifest in very basic parts of our getting around in the world. None of that history will be inherited by constructed machine agents.

And yet, we can tell similar stories about how artificial agents are constrained and how their constraints may well open up new ways of acting and perhaps being free. Return now to a fictional example for a moment to see how that might play out. Ava the android has no bodily features of her own as such. She does not develop from a cluster of cells executing a genetic algorithm in the environmental context of a womb. On the other hand, her capacity for information processing is as rich as, or richer than, our own. Her architecture is such that many different body parts, many different facial appearances and so forth are available to her. Her maker, Nathan, selects *one* particular example of a female appearance and binds her to it. She is then externally constrained to a single appearance. The important upshot of this appearance and its fixedness, however, is that it opens up possibilities for her given her representational and computational capacities. Remember that her task is to recruit Caleb as a co-conspirator to help her escape from her room and from Nathan's control. By picking for her Caleb's exact romantic/sexual type, Nathan gives her certain kinds of possibility that she would not have had as a disembodied machine mind, a robot with an industrial machinery appearance, or even a humanoid robot that was not attractive to Caleb. (Just this kind of worry is what prompted Turing in his devising of the imitation game to make sure the players were disembodied, so that no extraneous possibilities for cheating arose.) It is crucial to the execution of her plan that Caleb thinks of her as a victim of the abusing Nathan and that she needs rescuing. Her constraint to a single appearance opens up an entire range of choices for her. Even though this is a fictional example, we can see that Ava has a robust agency: her capacity to act is a product of her constraints and the possibilities those constraints generate.

IV. Final Word

Even if Ava can process information as well as we can or better, why are we so convinced that information processing has a lot to tell us about agency? Recall that Searle's so-called Chinese Room argument concedes that computers can process information. His insistence was that no amount of information processing itself could generate intentionality, something he and many philosophers have taken to be central to agency. Juarrero does not focus on that, but instead tries to illustrate the fundamental role of biology in explaining agency. While their emphases are very different, both Searle and Juarrero are focused on the fact that the capacities that machines have do not arise from processes like evolution, which we know can generate agency. It seems that the burden of proof rests on those, like us, who think that information processing can do all the things that are necessary for genuine agency. While we do not have settled answers here, we take up that challenge in the next chapter. The story will involve seeing information not as something to be abstractly processed, but rather as the foundation for how all agents, biological and artificial, are embedded in worldy contexts.

Your Tasks

Test Your Understanding
1. What is a complex dynamical system?
2. Why does Juarrero think that biology matters for agency?
3. What is an intention, according to Juarrero?

Reflect or Discuss

1. How is Juarrero's view of agency distinct from the views of agency we saw earlier in the book? How would you compare and contrast her account with the standard account that has the belief–desire pair at the center, or with the accounts offered by Frankfurt and Dennett?
2. Imagine you endorsed the account presented in this chapter. How might you view mindedness? Would you think brain states are the same as mental states? What about consciousness?
3. Juarrero writes that constraints are both freedom enhancing and freedom limiting. This is a difficult concept to get one's head around. So, here is one more example that helps. The Russian language has an extensive case system, with nouns, pronouns, adjectives, and determiners all inflecting (usually by means of different suffixes) to indicate their case. This allows for far greater flexibility in word order than in English, such that multiple configurations of one sentence are possible in Russian. Alexander Pushkin and William Shakespeare wrote poetry in different languages and therefore within different systems of constraint. The options available and unavailable to them within their respective systems (Russian and English) bounded as well as enhanced their creative freedom. It is only within these sets of constraints that innovation and novelty are possible. Now, can you think of another example from your own life or work that helps make this point?

Expand Your Thinking

1. Read Greg Egan's short story "Crystal Nights". How might Juarrero read this story? Would the artificial entities in the story count as biological agents on her account? Also, what are some of the ethical issues that arise in the story, and what might they reveal to us about the challenges of creating artificial entities?

2. Cordyceps—also called "zombie-ant fungus"—is a fungus that infects insects such as ants or spiders. The fungus drains its host completely of nutrients before filling the host's body with spores that will let the fungus reproduce. The fungus then *compels* the insect to seek height and remain there before the fungus expels these spores (thereby infecting other insects). Perhaps most shockingly, the fungus doesn't affect the insect's brain, and the insect remains fully aware of itself, even as it acts according to the fungus's will. What do you think this example tells us about agency and intentions? Is there anything else in this example that you find interesting or challenging, or that disrupts your view of mindedness and agency?

3. Consider a thought experiment: imagine that some material functionally equivalent to neurons has been discovered, and that surgery has become so precise that any given neuron could be extracted and replaced by a synthetic one. Would we think that having one synthetic neuron out of some eighty-six million organic neurons would change whether some person was an agent of the same sort as before? Would things change after replacing two neurons or even three? What, if anything, does the thought experiment tell us about whether a biological substrate is necessary for agency?

Further Reading

Denning, Peter, and Tim Bell. "The Information Paradox." *American Scientist* 100, no. 6 (2012): 470–477.

Deacon, Terrence. "On Human (Symbolic) Nature: How the Word Became Flesh." In *Embodiment in Evolution and Culture*, edited by Gregor Etzelmüller and Christian Tewes, 129–150. Mohr Siebeck, 2016.

Gopnik, Alison. "AI versus 4-Year-Olds." In *Possible Minds: 25 Ways of Looking at AI*, edited by John Brockman, 219–230. Penguin Press, 2019.

Juarrero, Alicia. *Dynamics in Action*. Boston: MIT Press, 1999.

9 Information, Communication, and Control

Introduction

Whatever one thinks of Searle's argument against the possibility of programming intentionality into computers—and as you may have gathered, at least one of your authors does not think much of it—the idea that one can make intentionality out of pure syntax does strikes us as wrong. We have said a little bit, much earlier, about why we think that computers are not best understood as operating with pure syntax. Even so, there are good reasons to think that bare computational capacities are insufficient, even where, for technical reasons, they go beyond pure syntax. The fact that all implementation is by means of causal powers does not by itself tell us that those causal powers are sufficient for intentionality ("aboutness" about the world), for reference, for meaning. In the chapter on computation, we defended the computational theory of agency but never quite settled what more than computation was necessary to make it work. Rather than leave things there, we want to say a little more about how it is that the artificially intelligent systems we are building can get in touch with the world in the right way, a way that gives them their own version of intentionality. They do it by gathering information about the world. And the thing is, that will turn out to be just how we do it as well, at least according to some prominent thinkers.

The capacity to get information about the world, to share that information, to operate in a variety of ways depending on what that information indicates about the world is all described under the umbrella of what is called "communication theory." (Sometimes, folks also call it "information theory," and it is sufficiently tightly connected to another discipline, cybernetics, that we will say something about that as well.) What was missing in the original so-called Chinese Room situation is any connection between the input signals, the output signals, and the worldly situation outside the room. That is also what is missing in the situation described in the Octopus argument. Whether one needs to have the *feeling* of understanding or whatever it is Searle thinks one lacks, once those signals are properly connected to the worldly situation, the so-called Chinese Room *as a whole*, including its occupant, does do everything a Chinese speaker does, and once the connection between inputs and outputs is properly structured, it can do everything that is normally involved in navigating a space where Chinese is the predominant language. But how does that connection get going? How does the ability to communicate translate into the power to act?

I. Command and Control

You might well have thought that communication had little to do with action. At least on the surface, communication is about sending and receiving messages and little to do with representation-guided behaviors. But here's the thing: with a sophisticated account of communication, the connection between communication and action is clearer because it makes more precise both the way representations are grounded in the world and the constraints the world imposes on agents' capacities to update their representations. It is crucial for action that behavior-guiding representations are constantly being updated in light of the way the world is, and the only way to find out how the world has changed or not during the course of executing some behavior is for information about the state of the world to be communicated to the agent.

Think back to Medea, the witch, in her interactions with Talos, the bronze statue Hephaestus created to protect Crete. Our favorite accounts of how Talos was defeated involve Medea *talking* to him. She convinces him to open his ankle and let out his own ichor. She does this by representing the world in a certain way, by getting *him* to represent the state where his ichor is drained out as one where his lot is somehow improved. This is just a story, of course, but it contains important lessons about what is possible using just our intuitive ideas about communication. The principal takeaway is that communication is a kind of *action*; it is action that can affect the behaviors of others by affecting their representations of the world.

We connect the idea of agency with doing things in the world, and we connect doing things in the world with a very general notion of communication—a notion articulated in a very clear way by Claude Shannon, the architect of the contemporary theory of information. In a way, the present chapter grounds our working theory of agency itself in communication theory. This helps both to clarify further our working theory of agency and to embed the account in the architecture of more interesting agents than the thermostat. "More interesting" in this context means that they can take advantage of wider communication channels: there are more signals that they can decode, and they have much larger libraries of messages. These are both meant to dovetail with the minimalist account of agency we saw earlier.

Shannon's communication theory is just one of two important strands of effort to understand how signals generated at one location could control the behaviors of things at distant locations. More and more devices such as rockets, probes, airplanes, and a whole variety of other equipment were being remotely controlled, and an analytical account of the limits and scope of such control was needed. Norbert Wiener is the most prominent figure associated with the development of the second strand, what he called "cybernetics," after the Greek word *kubernetes* (meaning "the person who steers the ship").

Wiener's idea was to explore the limits of how to use various signals (electronic, acoustic, visual, etc.) to control the behavior of the systems that receive them. These systems are not confined to machines but rather include all sorts of other things, even animals, and even us. Much of the branch of study relevant to our discussion of self-driving cars and other kinds of robots is rooted in cybernetics. As far as an analysis of agency is concerned, the fundamental results of that field coincide with those of communication theory. We opt, however, to focus on communication theory for two reasons. First, that theory is much more closely connected historically to computational issues and the representational capacities of computing devices, and computation is where much of the conversation about AI is centered. Second, communication theory is tightly connected to our way of thinking about agency as behavior guided by representations. The connection between agency, representation, and signaling seems to us easier to draw using communication theory than cybernetics. Again, we do not claim any fundamental advantage of communication theory over cybernetics; our decision is a practical one.

Communication covers a lot of ground. Normally, it seems, when one thinks about communication, the focus is simply on the exchange of information, and information is thought of more or less inertly. It is easy to think that because the felt causal power of some information is rather slight, so too is its actual causal power. But consider that the feeling in my eyes of seeing clouds rolling in, signaling rain, is barely different from the feeling in my eyes of seeing clouds rolling out, signaling the return of nice weather. At the same time, there is a huge difference in the implications of these different signals for my action, for the impact of that action on the world at large. We are, to a greater or lesser extent in various circumstances, controlled by the signals that we receive from the world. Understanding precisely how that works is the business of communication theory.

II. Information Gathering and Utilizing Systems

Communication theory is a theory about how information flows and how constraints on that flow make it possible, when it is possible, to solve what is called the "effectiveness problem." That problem is that sometimes we want to affect the world without walking over to where we want the effect to happen and making it happen. Sometimes, when we want to affect the world in certain ways, we can only do so indirectly by communicating our goals. But even with a willing collaborator (a friend, an ally, Talos even), there is a limit to how much and how well we can communicate what we want. A measure of how well we do that is the connection between our messaging and the change in the world—that is, how *effective* our communication is. Information is a measure of how much we communicate, and the theory of information flow is a theory about how much it is *possible* to communicate, given a particular channel of communication (a telephone wire, spoken word, printed text, visual signals, etc.).

What, exactly, *is* information, how does it flow, and what is it to *process* it? It seems that we all have at least an intuitive notion of what it is to be informed, but beyond that, the concept of information seems vague. Let's open with this: for anyone or anything to be informed of something is for some signal to carry information about that something to it.

What is it for a signal to bear information about something (some situation, fact, or what have you)? At its simplest, it is for a lack of certainty about something to be transformed to certainty by means of that signal. Here's an example. Suppose we're playing cards, and you know that I have either the ace of clubs or the two of hearts in my hand (however it is you might know that). If you draw the ace of clubs or someone else we're playing with discards it, then whichever happens, this *signals* to you that in fact I have the two of hearts. Your uncertainty is replaced with certainty. You have been informed about the card I have; the information about that card has flowed to you. That reflects our standard usage pretty well. We say, "The discarded ace of clubs signaled that, in fact, I had the two of hearts in my hand."

In general, signals can be of many types and of disparate natures. My standing around somewhere, the sparkling lights in the Eiffel tower being on, the ding of an elevator arriving are all signals that, for the right agent, can be seen as signals that bear interesting pieces of information: that I want the bus to stop to pick me up, that it's less than five minutes after the hour during the night in Paris, that the assassin has arrived and I need to run. We human agents are capable of receiving tremendous amounts of information via our sensory apparatus. Our eyes, ears, nose, tongue, and skin permit many signals of many different kinds to bring us massive amounts of information about the world all at once. Other agents are more limited in their information-receiving capacities. A clam or an amoeba seems to have only one kind of input and only a little at a time.

Let's now hook this into our working theory of agency, where to act is to behave for the sake of some end by continuing various behaviors until a representation of the world matches a set end state (within some tolerance). Limitations on information flow put limitations on representational resources and their appropriate updating. Moreover, a system's capacity to deploy elements of its repertoire of actions effectively is also constrained by the information it can have about the world. Limitations on information input, restriction on its flow, amount to limitations on the agency of various entities. With this background, let's define and quantify information flow.

You might well wonder why we are not defining information itself. That's because, on our view, information is what physicists would call a "bad physical quantity." There are, first, competing definitions of it. Moreover, each of them suffers from a kind of ambiguity and fails to transform under changes of description the way good physical quantities do (quantities such as momentum or energy). In this respect, information is like heat and total gravitational energy: these quantities are not well defined and are not good physical quantities, and yet their *flow* is well defined and a good physical quantity. Moreover, the flow of information is the real quantity of interest; information itself does not do any work for us.

Shannon spent a lot of time figuring out how best to design telephone transmission networks, and he posed and answered an important question: given a channel in which information flows, just how much information can it carry? Implicit in this question is the conviction that information does flow in channels. But what does that mean? Just this: various kinds of signals are transmitted by means of various kinds of media—cable television signals in coaxial cables, cellular telephone signals via electromagnetic waves, nineteenth-century mail via networks of ponies and carriers, and so forth. Each of these media counts as a channel for the flow of information, and the signals are what bear that information.

In addition to sending and receiving signals, communication requires that these signals be related in the right way to the message that is being communicated and that the signal can be decoded properly to recover that message. If we speak to English speakers in French, then the spoken signals may well be related properly to our message, but they may not successfully communicate that message. One needs to be able to understand—that is, decode—the signal. This decoded signal must pick out one of a library of possible messages that the receiver and the sender share in advance. In the simplest case, there might be only one message. For example, at a silent auction, the message is "I bid that much," and the signal is the raised paddle with my number on it. In other cases, like spoken communication, we have stores of concepts that, as children, we are raised to associate with various words. In later life, hearing a word and bringing to mind the concept is our receiving the signal, decoding it, and picking out the message that was sent.

Think of some signals used by a sports ball team: there are very short phrases that are linked to very long sequences of actions and contingent subactions, given what happens during the next few moments of the game. The coach, earlier during training, has outlined these sequences and indicated the code for them. When you hear the team leader shout out "raspberry 7," for example, you and your teammates select the appropriate routine to execute.

Given this huge array of possible types of signal, what hope do we have for quantifying the flow in general? Shannon suggests that information flow is related to how we go about reducing our uncertainty about what the transmitter of that information is attempting to inform us about. Instead of focusing on the types of signals themselves, we begin with the idea of a store of possible messages.

Consider the easy case where there are only two possible messages of interest. Remember what happened on Paul Revere's ride: his task was to clarify whether the British were attacking overland or by ship. "One if by land, two if by sea" is a slogan that tells us how to decode the signal Paul carried: if the British were coming by land, he would carry one lantern; if they were coming by ship, he would carry two. This works because his audience already knew the possible messages and that there were only two distinct signals. Uncertain about the intentions of the British, they relied on the number of lanterns to disambiguate between them. The slogan in this case defines a particular channel in which information flows—what Shannon called the "channel of communication" because, for him, information is all about communication. And the thing is, given that channel, no more information can flow than is required to choose between those two possibilities. If the British could have been flying in on helicopters, then the message scheme would not have worked. We would have needed a more capacious channel—perhaps light on the left if by land, light on the right if by air, and two lights if by sea. This scheme would not lend itself so well to a dramatic poem about the invasion as did the original, but it would allow for more information to be sent—that is, for more possible messages to be chosen among.

Generalizing this insight, Shannon suggests that the unit of information cuts in half the number of possible messages we need to choose among. Suppose that you and I agree in advance that the book we will read for our study group next week is one of two that we have been considering. In that case, we could make a signaling scheme with only two options in order for one of us to convey the choice to the other. But if there were four books, say, we could get by with a scheme that, first, cut the list in half, and then cut it in half again—that is, we could use a two-option scheme followed by another two-option scheme and so on. Each doubling of the number of possibilities requires one more binary choice. It's a mathematical fact of some interest that this scheme is neatly captured by relating the quantity of transmitted information to the base 2 logarithm of the number of possible messages. It is *arbitrary* that our scheme goes like that, but it is also in some sense natural to choose it. That captures a key insight of Shannon's theory for our purposes— the idea that information flow is about the reduction of uncertainty. There's a lot more to explore in communication theory that is beyond our scope here, but this gives us the right footing.

III.   Updating Representations

We have been speaking as though communication theory and information theory are more or less the same thing. We have been talking about communication as encoding and decoding signals to identify a message from a shared library of messages. What could that have to do with observing the world itself? When I try to communicate with you, and we have a shared library of possible books, or messages, or whatever that we are choosing between, it seems to have little to do with my getting information about the world by means of my senses; I am just getting information about what you want me to know. The world, by contrast, does not *want* me to know anything, and there is no library of messages that the world may be sending me that I am picking through when I use my senses to learn about the world. So, what is going on there? What connects Shannon's particular interest in stores of messages with our more general attempts to find out what is happening out in the world?

Here's one way to think about things. Consider the kinds of "yes" or "no" questions you can ask about the world. Is this object heavier than that one? Does this electron go up after passing by this magnet? Is the trajectory of this particle as my theory predicts? Given questions of this sort, the world will answer them for you, in many cases (in those cases when you have the observational resources to hear, or see, or feel that answer). Here, then, the store of possible messages is a simple one. It is the "yes" or "no" response to your question. That tells us that even though there is no list of books, say, that we are picking up on cue from a given signal, we can still use Shannon's story to understand sensory apparatuses as receiving information-bearing signals from the world.

Shannon's account tells us how much information can flow in a given channel, like a fiber-optic cable. It tells us how to optimize our coding schemes to maximize the carrying capacity of signals in a channel. But it does not tell us much about whether information will actually flow at the end of the day. Partly, that is because whether or not information flows has to do with what information agents at the receiving end already have. You cannot receive information that you already have. Suppose someone asks

you to go get a book off their shelf but does not say which one. You may
wonder what book it is that they want. If you get a message from them
telling you which book, then you are now informed of their choice. But if
they then send *another* message telling you the same thing, you do not get
informed again. If anything, you get annoyed. "I heard you the first time,"
you might mutter. Being informed once is all there is to being informed.
Information flow, then, relies on the state of uncertainty of the receiver of
a signal. Notice that if you already know which book to choose and you are
with someone who does not, then the two of you may well receive the exact
same signal, and your companion receives information while you do not. We
are constantly bombarded by signals of various sorts, many of which are in
information-bearing channels, but not all of those convey information to us
either because we cannot decode the signal or because we already know what
is being conveyed.

Receiving information requires both that one be connected to the proper
signals and that those signals suffice to disambiguate matters. That latter
requirement itself has some sub-requirements. We saw the first already: there
must be an ambiguity to resolve—the agent must be entertaining at least two
different possibilities with no way to select among them. Another is this:
the agent must be sufficiently sophisticated to take advantage of the signal
to resolve that ambiguity—that is, it must be capable of processing the signal in order to put itself into a position to update its appraisal of the situation. In fact, typically something like this is what is meant when people talk
about "information processing." Information is not the kind of thing that
gets processed, but receiving information is predicated on changing the state
of uncertainty of whoever receives it.

IV. Agency and Information

At this point, you might have lost track of how all of this message and signal
talk is related to the concerns of artificial agents, or agents in general, who
want to affect things. What, again, does communication have to do with
agency? Warren Weaver puts it very well:

> The word communication will be used here in a very broad sense to include all of
> the procedures by which one mind may affect another. This, of course, involves not
> only written and oral speech, but also music, the pictorial arts, the theater, the bal-
> let, and in fact all human behavior. In some connections it may be desirable to use

a still broader definition of communication, namely, one which would include the procedures by means of which one mechanism (say automatic equipment to track an airplane and to compute its probable future positions) affects another mechanism (say a guided missile chasing this airplane). (Shannon and Weaver 1963, 1)

Weaver refers to minds affecting each other. How so? Well, one of us might believe something—say, that the British are coming by land—and want the rest of us to believe it too. The procedure that is used is the swinging of one lantern. That induces in us the belief that the British are coming by land. But, as he says, this can be more general. One direction of generalization is from belief to representation. A more general agent's representation of the world can be updated by a signal sent by another agent or even by the world itself. And that change of representation changes its possibilities for action. Communication theory has a long reach. When we think of behavior as Dretske does, as outward motions brought about by internal causes, then one system affecting another's behavior is anything that it does that prompts that other system to change any of its motions in the light of the signals the first one sends.

Remember, agents intervene in the world; they need to make changes to bring their representations into alignment. They make a change, register that change, and recalculate the divergence between their end-state and present-state representations. They then adjust their behavior in that light. All of this falls under what Weaver calls the extended sense of communication by which one system influences the behavior of another—even though, here, some of that is under the guise of one part of one system (a subsystem) influencing the behavior of another part of that same system.

Consider the example of a map app on your phone. This app involves a representation of the phone's location and its current trajectory. It can compare the representation of the phone's location and its current trajectory with a representation of the selected path. It has the capacity to speak (behavior), instructing the user of the phone to modify or not the trajectory of the phone. It continues to issue these instructions until the representation of the phone location coincides with the representation of the selected end point. It is in communication with the world by means of various signals (cell, Wi-Fi, etc.). It decodes those signals and identifies appropriate elements of its store of messages: off path, on path, and so on. It also picks out elements of its store of messages to encode and communicate to its users. These behaviors, guided as they are by its representations, are actions.

Acting, as we have seen, at least involves changing the world—either by the effect we have on our surroundings or, sometimes, by changing *ourselves* to align better with the world. At bottom, that is the subject of Shannon's theory of communication: to determine the bounds of our influence, and others on us, given the physics of the situation in which we exist. The same goes for all agents who deploy some elements of a behavioral repertoire until the world-state representation matches the end-state representation.

Shannon's radical idea was that the store of messages, the coded signal, the receiver, the decoded signal, the selected message itself are all part and parcel of communication, and therefore all talking, seeing, hearing, gesturing, and so on is constrained by the capacities of information-bearing channels. But as Weaver notes, this can be extended to an account of any way that signals change representations and thus the behaviors that they guide. It thus applies to all sorts of agents.

We can now look back and relate this account of information to one of Juarrero's critiques of information theory as an account of agency. Juarrero suggests that information theory comes closer to explaining the processes by which intentions are manifested than classical mechanical theories but claims that it too is inadequate. Information theory, she acknowledges, provides a mathematical characterization of the relationship between the sender system and the receiver system (Juarrero 1999, 98). The technical concepts of noise and equivocation provide us with the means of measuring and understanding how information can flow without interruption from source to terminus, which Newtonian causality, she thinks, could not. However, to understand biological systems such as human organisms, she also thinks we need something more than a content-free characterization of information— we need semantically relevant information. On Juarrero's view, information theory cannot explain why it is that a rat, say, can learn to avoid food that makes it sick after one instance, but it will take a month of conditioning to learn to avoid food that is accompanied by a shock. To explain that, says Juarrero, we need a story about the kinds of endogenous constraints that are present in the rat as a learning system, as well as where those constraints come from. This means embedding the rat in a history and in an ecological situation where the quality of food matters. Where information theory fails, complex adaptive systems theory steps in, Juarrero thinks.

This, on our view, reflects a good understanding of the technical parts of information flow but not the key point about the shared library. Rats are, as one might say, constantly probing the world for the answer to "Is this good to eat?" and the way the world bears information to them about that is by either making them sick or not. Rats have, so to speak, an ambiguity in mind that needs resolving in light of some signal from the world. They have no similar communication channel set up that is coded to electric shocks. So, part of their training must first be to set that channel up as a way to resolve certain ambiguities such as, "Is the floor safe to step on?" and, over time, "Is that food good to eat?" Now, Juarrero is quite right that information theory itself does not say anything about which channels and libraries and signal-uptake systems rats will have. Nor, by the same token, does it say anything about how copper wires or cell-phone towers work. None of that is to say, however, that agency as such is interestingly connected with complex adaptive systems in a way that is opaque to illumination by information theory. Rather, it is a contingent fact about various systems what their channels of information flow are and whether some of them come built-in (e.g., by evolution or by a roboticist) with connections to libraries of decision versus needing to be built on the fly.

Here's a potential rebuttal Juarrero could pose. If we want to reverse engineer the rat's responses, we could make a machine that processes information in just the right way to mimic what the rat does in the case of shocks and in the case of bad food. But the rat is not a computer. It expresses its agency by means of its complex multilevel causal structure. And that is a complicated evolutionarily designed structure.

This is a strong rebuttal, and we could go back and forth with Juarrero. For example, we could respond to the rebuttal by saying that yes, that may be so, but as long as we agree that information theory suffices to describe what is going on, we do not have to say that we ourselves could design things the way evolution has. It has had millions of years to design these very sophisticated information channels. That is how organisms implement agency. It may not be how machines do. To this, Juarrero could respond with an additional point. Much of philosophy is going back and forth, offering arguments and counterarguments and rebuttals, as we all collectively contribute to the well of human knowledge and understanding.

V. Putting It All Together

We began this book talking about the way artificial agents have been portrayed in fiction in the West, starting as early as Homer. Those depictions, and the thought that many technologists were working on building more sophisticated machines, prompted a discussion of the nature of agency itself, and we developed a number of philosophical tools to analyze and clarify the concept of agency. While we considered a number of approaches, we finally adopted a minimalist conception—one that seems at least promising as an account of agency that could be suited to describing the most basic of machine agents and then be further elaborated to cover more and more sophisticated machines.

We then spent time saying more about what machine agents could be like, but more centrally to an examination of the conditions that would have to be in place before the agency of machines would be recognizable as of the same sort of thing as the kind of agency we take for granted in even very simple natural agents. These conditions amounted to having the capacity first to represent various states of systems in the world using states of the machine itself and to use such representations to control the deployment of a behavioral repertoire itself capable of modifying the states of those worldly systems until their representations match that of some end-state representation. We presented these requirements through the lens of communication theory and the theory of computation, and we also considered competing but compelling accounts from a more biology-focused perspective.

There is much more to say about the fundamental conception of agency itself; these conceptual issues are far from resolved. Equally important are the ethical issues that arise from the conceptual points we have been covering in the book. And it is to an examination of some of those that we now turn. We focus on responsibility, moral status, and relationships, but there is a long list of potential ethical topics to dive into. What follows is more questioning, more exploration, and we hope it will spark in you an interest to ask additional ethical questions about what increasingly sophisticated machines will mean for our moral lives.

Your Tasks

Test Your Understanding
1. Define "information" and "signal" and come up with one original example for each.
2. Summarize Shannon's communication theory in your own words.
3. Explain, in your own words, the relationship between information and agency as presented in this chapter.

Reflect or Discuss

1. How might Juarrero respond to the objections raised against her view of agency? Develop the strongest possible counterargument that you can on her behalf.

2. Can you be an agent without being an information processor? That is, is it possible to be an agent without having some capacity for information processing?

3. Evaluate Shannon's theory. What are it's strengths and weaknesses?

Expand Your Thinking

1. Look back at the Talos and Medea story, as well as to the Octopus test and so-called Chinese Room thought experiment. In all of these, language is taken to be important. What role(s) does language play in the examples? How might, if at all, language fit into a theory of agency?

2. Both this chapter on information and the chapter on the computational view of agency provided ways of deepening the minimalist theory of agency. Looking at these chapters together, are there any tensions or gaps or problems that you noticed? What about redundancies? If so, reflect on these weaknesses of the view and consider ways they can be improved.

3. Many scholars have noted Shannon's importance in the origins of large language models. Do some digging into that history. Then, reflect on what Shannon might say about LLMs and the other machines we have talked about in the book.

Further Reading

Dretske, Fred I. "Precis of Knowledge and the Flow of Information." *The Behavioral and Brains Sciences* 6 (1983): 55–90.

Juarrero, Alicia. *Dynamics in Action*. Boston: MIT Press, 1999.

Mitchell, M. and Krakauer, D. C. "The debate over understanding in AI's large language models." *Proceedings of the National Academy of Sciences*, 120, no. 13 (2023).

Mattingly, James. *Information and Experimental Knowledge*. Chicago: Chicago University Press, 2021.

Shannon, Claude E., and Warren Weaver. *The Mathematical Theory of Communication*. Champaign: University of Illinois Press, 1963 [1949].

10 Responsibility for Machine Actions

Introduction

We have discussed a large number of features of machine agency and agency generally. We have considered a lot of questions: What grounds agency in all of these cases? What features does agency have, and what features should an account of agency have? Which entities in the world are agents? What would it take to make an agent from scratch? And while we have given our working account of agency, we have asked you to consider these things independently of any particular account. One thing, indeed, that does seem to transcend particular theories of agency is that sometimes agents, when they do things, when they act, are responsible for the things they do.

In philosophy, theories of responsibility have focused on the agency of the individual and asked whether a given individual meets the criteria for moral responsibility. These conditions have generally included control and knowledge. Not only has the theme of responsibility been explored in moral philosophy, but in metaphysics as well, where topics such as free will and determinism have put responsibility at the center of discussion.

Agency tends to go hand in hand with responsibility, but the possibility of machine agency raises difficult questions. Some folks are concerned about the fact that machines will be performing more and more actions with significant consequences, coupled with the further fact that the agency of the machine is much more extensive and so those actions will be further and further removed from the control of any human. The worry is, were this impact caused by a human, it would require the human to be held responsible, and yet there is no human in the picture that could reasonably be said to be responsible. This is one way of describing the so-called responsibility gap that seems to be opening as the result of the increasing capacities of agential machines: human creations are increasingly doing things that no human is well positioned to take responsibility for. In this chapter, we will try to figure out whether, and if so, how, we can close or bridge this gap.

I. What Is Responsibility?

Humans are interesting creatures. When they begin life, they are not the kind of entity that can be responsible for much. Their main causal power seems to consist in keeping parents awake all night. Over time, however, they become causally more powerful and are responsible for more of the things that go on in the world around them: they make bigger messes, they snatch at things, they break dishes, and they speak . . . a lot. And yet even then, when they are *causally* responsible for a lot of different effects, they generally cannot be *held* responsible for the things they do. Only as their agency increases do they become ever more subject to being held responsible for their purposive behaviors, for their actions.

Children do not become responsible agents in a vacuum, however. In paradigmatic cases, there is a reciprocal process where children are *given* more responsibility and can thereby be held responsible for a new range of behaviors. Of course, our society does not always function this way, and the link between being given responsibility and being held responsible is messy and unclear. There remains, however, some conceptual link between these two aspects of responsibility.

What does it mean to be responsible for something? Let's look at different but related ways of assessing this. First, there is a kind of standard difference between causal responsibility versus moral responsibility versus legal responsibility that you are probably familiar with, even if it's never been explained to you in these terms. Here's an example: Arjun falls and hurts his knee. Why did he fall? Because a part of Tyler's body, her hands, collided with Arjun's back, causing Arjun to fall and hurt his knee. We would generally say that Tyler is causally responsible for Arjun falling. That's the barest beginning.

The next thing to wonder about is, What happened? We do not know enough from the listing of various behaviors and happenings because action is not just behavior, remember. It is behavior *under* a description, or *for* reasons, or *guided by* representations. So, to know what the act is, we need to know what the right description of the situation is. Did Arjun fall because he was an actor in a play playing a part? If so, that would render the kind of action different than if he weren't. Moreover, whether Tyler is morally responsible for Arjun falling will depend in part on whether she acted intentionally or accidentally and the context in which the action took place. Here is where our minimalism about representations should be given another hard look. Tyler's intention is doing a lot of work here, and it's intention understood as "desire for some outcome" rather than the more stripped-down version of intention understood as "aboutness." Can an agent that only has the latter possibly be put into a context of moral responsibility?

The fact is, though, that we cannot determine moral responsibility from the information we have so far. And that is to be expected. Similarly, we need more information in order to determine whether Tyler can be held legally responsible. It depends on who Tyler is. If Tyler and Arjun are both five years old and in the playground, then even if Tyler intentionally pushed Arjun, she is probably not legally responsible for the action in any meaningful way (unless the injury sustained were very serious). But the situation looks different if Arjun and Tyler are both adults, or there is an age difference, or if Tyler is acting in a diminished capacity.

In addition to asking who the agent is in a single-glance kind of way, it is also worth considering the idea that there can be different degrees or classes of responsibility, and that these may be dependent not merely on who one is but also on what roles one has taken on or the relationships one is in. People in general are not responsible for certain things that those occupying particular roles might be; doctors, lawyers, and cops are responsible in ways specific to their roles when they act in that role's capacity. Additionally, our responsibility toward our family, our friends, our neighbors, our employees, and our fellow citizens and other humans are of various different sorts. One simply cannot assess responsibility without knowing a lot about the context that makes the act what it is, who performed the act, and those affected by the act.

In brief, responsibility is complicated. There are things, for example, that are simply causally responsible for various happenings (the sun heats the earth, and rivers flood towns) and cannot be held responsible. But agents, when they are responsible, are subject to being *held* responsible: their doing of a thing sometimes entails that they incur debts or obligations whether they be financial or legal or moral; sometimes, it entails that they gain entitlements as well.

Everything gets a bit muddier when we explore the possibility of holding machines responsible for their actions.

II. The Responsibility Gap

In philosophy, we can make both descriptive claims and normative claims. Descriptive claims tell us something about how the world is in fact. *"Frankenstein* is a novel written by Mary Shelley" is a descriptive claim, as is "Some people believe that when a person dies, their soul goes to heaven." Normative claims tell us something about how the world should be. "We should not lie" and "I ought to read more" are both normative claims. Turning back to our stories, claims such as "Ava should not have left Caleb to die" and "Frankenstein's Creature should not have killed anyone" are both normative claims. This distinction here is not the difference between moral and nonmoral. Moral questions are often based on both descriptive claims and normative ones.

On the issue of machine responsibility, there is both a descriptive and normative component. The descriptive question is, If machines can be agents, does that imply that they are responsible for their actions? The normative question is, Who should we hold responsible for machines' actions?

Let's consider a specific example. Self-driving cars are now a familiar feature of the world. Perhaps many of us have not seen self-driving cars on the road, but we know about them. Self-driving cars are not all the same: there is a wide range of autonomy. Some of these vehicles have just the capacity for emergency braking to avoid crashes. Others have full autonomy. For example, a location can be set, and then the car can drive off to its destination without any humans at all inside. Not only does that technology exist, but in some states, it is legal for people to "operate" vehicles simply by being the person who programmed in their destinations.

Self-driving cars are every bit as capable of causing harm as those driven by humans, even if their probability of causing that harm is lower. These cars come programmed with assessment routines installed; these guide the car to one option among many when threats to life and property are detected. When things go badly, and someone or something is harmed by the moving car, who is responsible? Is it the "driver" who may be entirely passive in that position in the near future? Is it the manufacturer who installed those routines, and are they only responsible when the routines are badly implemented or also when they are functioning "correctly" but we think they're the wrong routines? Is it the ethicist who advised on the selection of the assessment routines? The example of the self-driving car, and the difficulty we have in attributing responsibly neatly, calls to mind an important literature related to the so-called responsibility gap.

The responsibility gap, first introduced by Andreas Matthias, refers to a new type of problem raised by the increasing class of machine actions for which we cannot use our traditional ways of responsibility ascription because "nobody has enough control over the machine's actions to be able to assume the responsibility for them" (Matthias 2004, 177). Determining who should be held responsible for the actions of these machines is the core issue to be resolved. Could it be the machine?

It may strike some ears as funny to speak about machines being responsible. Many have thought that it would just be a category mistake—machines are just not the right kind of thing *to be* responsible. Deborah Johnson, a prominent thinker in the field, does not believe machines meet the requirement for moral agency: they do not have mental states, and even if states of machines could be construed as mental states, these machines "do not have intendings to act arising from their freedom" (Johnson 2014, 20). Relatedly, Patrick Chisan Hew writes that thermostats, mousetraps, toilet tank-fill valves, and automobile cruise controls are all examples not only of agents but also of intelligent agents (albeit simple ones; Hew 2014, 198). He accepts that an agent is something that can act in the world. But being an agent is not sufficient for being a moral agent, on Hew's view: "an agent is *morally praiseworthy*, and can be held *morally responsible* for an action, if it is worthy of praise for having performed the action. We call such an agent a *moral agent*" (Hew 2014, 198). It would be odd, Hew says, to call a mousetrap a moral agent because the rules for its behavior and the mechanisms for supplying those rules are determined by external humans. Responsibility for the mousetrap is held by humans because, from "trigger to trap," humans are the ones who "armed it, designed it, constructed it and so on" (Hew 2014, 198). For Hew, simple agents such as mousetraps and thermostats are not held to be moral agents. Similarly, Mark Coeckelberg (2009) writes that humans can ascribe virtual moral responsibility to those entities that appear similar to themselves. He also notes that assessing whether a machine is worthy of moral praise depends on whether the machine is a *free and conscious agent*, which we have no way of determining.

For thinkers such as Johnson, Hew, and Coeckelberg, people, and perhaps entities like people in the right ways, can be responsible. This seems to also be true in the emerging literature on AI agency focused on contemporary machine learning systems like LLMs, which tends to separate the question of which AI systems count as agents from the question of whether those systems are moral agents. Both Chan et al. and Shavit et al. stress that their exploration of agency, respectively, is not about moral patiency or agency. Carlsmith and Dung, meanwhile, acknowledge that their respective theories of agency might have further ethical implications, and suggest that agentic AI systems may be worthy of moral consideration, but they leave these as open and controversial questions.

But why is it hard to hold machines responsible?

You might think the problem is easily solved, given the tools we have been developing over the course of the book. Why not simply hold the machine responsible, if it is the agent, and be done with it? There are two broad, related difficulties with moving so quickly.

The first difficulty is that it is unclear, even given our discussion up to this point, whether and how we *can* hold machines responsible. We tend to think of holding responsible in terms that are related to human motivations. We praise and blame, we reward and punish. This, for prominent thinkers on moral responsibility, namely Peter Strawson, is at the heart of moral responsibility. Thinking in these terms, however, narrows the class of things we can imagine holding responsible to those things that are motivated the way that we, humans, are. Machines are set in motion by other forces than hope of reward or fear of punishment. Humans are as well, and we admire those who act from abstract principles (of justice, or fairness, or love). But much of our way of holding responsible still revolves around praise and blame, and that makes it hard to get machines with their opaque motivations into the picture. How can it be possible to hold machines responsible if we do not assume they will ever have the moral sensitivities that we do?

The second difficulty is that holding the machine responsible seems to make it too easy to let those humans who have built and deployed these machines off the hook. The idea is that we want Microsoft to be responsible for the technology it puts out into the world. Why? Because holding humans responsible is safe, and we know how to do it well. We have developed rich moral and legal systems that, while imperfect, are incredibly important for ensuring society functions. When we begin letting humans or groups of humans—especially powerful corporations—say that it is the machine, and not them, that is responsible, we provide new escape routes for humans and perhaps make ourselves less safe over time.

Let's say we want to ensure we do not provide cover for bad actors who will hide behind the actions of machines in order to avoid their own real responsibility for what they have done. Even if we were to see machines as responsible for various actions, we would still need to figure out how to continue to hold humans responsible as well, when appropriate. That is, we must distribute responsibility between machines and humans in such a way that we really are holding the machines *and* the humans responsible.

Scholars have responded to this concern in a number of ways. Some have explored legal avenues for holding machines directly responsible (Asaro 2007), while others have suggested that we can bridge the gap by allocating responsibility to both machine and human (Hanson 2009; Hellström 2013) or to a collection of humans in the creation of the machine (Marino and Tamburrini 2006; Rahwan 2018). Taking a more indirect route to addressing the gap, other scholars have responded to the gap by arguing we should program machines to be ethical (Arkin 2010). This is a move that Colin Allen and Wendell Wallach explore carefully in *Moral Machines*. Notice, however, that the idea of programming morality into machines implies that (1) we all agree about what morality entails, and (2) morality is the kind of thing that can be programmed. We can critique both of these premises. We leave these questions open and invite you to keep reflecting.

Other scholars, in response, have denied that there is a gap, (Tigard 2020), or challenged the framing of the gap, (Johnson 2014). Other scholars have similarly resisted this framing by denying one of the so-called responsibility gap's core assumptions: that control is a *prerequisite* for responsibility (Nagenborg et al. 2008; Santoro et al. 2008). Those in this camp point to other responsibility practices, such as holding individuals or corporate entities responsible despite the fact that they are not able to control the outcome. Thus, there is no problem with holding people responsible for the actions of machines, even if the people are not fully in control.

In the next section, we'll do a deep dive into a creative way of holding responsible, provided by Kate Darling, that does not rely on holding a machine agent directly responsible.

III. An Analogy to Animals

Kate Darling's monograph *New Breed* explores how we might create the right infrastructure for accommodating increasingly agential machines, with an eye to the conceptual, moral, and legal issues informing the debate on how we may keep humans in the loop. Darling's thesis is simple: by looking at humankind's history with animals, we will find inspiration for how we may incorporate future robots into our social world. Darling's account attempts to show how we can make sure that, even if we have increasingly autonomous machines, the buck always stops with the human.

Darling begins by challenging the assumption that it is apt to compare robots to humans. Looking elsewhere for an appropriate analogy, Darling suggests we should compare robots to animals. Darling is not arguing they are the same, but she does think that animals are a better reference point than humans in determining how we ought to regard future machines. On her view, using the animal-robot analogy ensures that we do not mistakenly assign responsibility to robots but continue to hold humans responsible, as we have in the case of animals.

To make her case, Darling begins by revealing to us how humans have used animals in ways that we today use machines. As late as 1870–1871, when Paris was under siege and all communication lines were cut, the French used the pigeon post system to bring in messages from outside the city. There are also darker cases. Before the advent of automatic guided missiles, the US Army tried to strap incendiary bombs onto bats, with the aim of releasing the "bat bombs" over cities in Japan. Darling's point is that animals and machines occupy more similar social roles than is often supposed or acknowledged.

Darling then explains the myriad ways humans have created laws to deal with "rule-breaking" animals. Both the Code of Eshnunna and the Code of Hammurabi, some of the earliest laws known to us (we do not know whether or how they were enforced), establish clear consequences for when animals cause harm. For example, both codes follow the same principle for oxen owners. If an ox owner knew the animal was a risk, the owner is penalized. But if the owner's ox unexpectedly kills someone, then the owner is not responsible. Biblical law, too, dealt with goring oxen, as did ancient Roman law, which introduced the concept of "noxal surrender"—handing over the problematic animal (or, generally, whatever dependent, including humans, had performed the harmful act). The ancient Romans also distinguished between domestic and wild animals, and dealt with special cases, such as bees, who present a tough edge case. More recently in history, other legal channels have been used for dealing with badly behaved animals. Pig trials were a common feature of the Middle Ages; during the Industrial Revolution, new rules were implemented for letting animals roam the streets; and most recently, in places such as Vienna, licenses are used for specific breeds of dogs that are more likely to cause harm.

This history shows us that humans, for a very long time, have come up with creative ways of dealing with challenges presented by the actions of autonomous agents who are not themselves legally responsible for their actions. Darling's core point is that we do not need to try to program morality into machines, nor must we ask robots themselves to make moral decisions. We should instead, she thinks, develop an architecture that holds humans responsible and mitigates harm to the greatest degree possible. Just as we have regulated animals and their owners in our legal system, so too can we regulate machines and their creators and manufacturers. According to Darling, there are many possibilities on the table. For example, we can develop context-specific rules for machines; rules for robots that roam the streets and skies in densely populated areas should be stricter and perhaps our priority. Moreover, we could introduce robot regulations that ameliorate harm that creators or users cannot anticipate, like ensuring "that [a robot's] physical build is made safer (the robot equivalent to muzzles), that their handlers are able to control them (likes the leashes and licenses mandated for dogs), and could consider a disclosure of risk (e.g., requiring signage on the property where the dog or robot is kept)" (Darling 2021, 74).

On Darling's account, companies that create and manufacture machines ought to be held responsible when things go wrong. Why? Because, Darling exclaims, "Incentives!" (Darling 2021, 84). Even if the chains of causality are complex, companies should be ultimately responsible for the technology they put out into the world so that they will think more carefully before releasing something potentially dangerous into the wild. Darling acknowledges that we are working against our biases. Often, when a human–robot interaction goes wrong (a Tesla or a plane crashes), we tend to deflect blame away from the machine and its creators and onto the operator, even if the human operator is not 100 percent at fault. But we ought, Darling insists, to resist this temptation. Holding the human makers of machines responsible for machine action is, on her view, the best option we have.

Darling offers a coherent and often compelling story—one that allows us to imagine robots as *"a different kind of agent"* (Darling 2021, xv, emphasis added). The question for us is, Does Darling give us the best picture for imaging a robot agent?

We need to scrutinize Darling's claims closely. Darling, for example, is right to suggest that no existing robots come close to exhibiting the artificial general intelligence that we fear in science fiction. One of the strengths of Darling's project is that she thinks more creatively about what future robots could look like. Notably, her case hinges on the notion that AGI will not exist. So, we do not have to think too hard about how we will incorporate AGI into our social, political, and legal infrastructure. But even if we do not develop AGI, it seems at least plausible that the kinds of robots we have in the future are not appropriately like animals with respect to intelligence or agency. For example, we could have long conversations with robots, which we cannot do with animals. Additionally, humankind's past history and current relationship with animals is often a brutal one, as Darling acknowledges. Darling needs to provide more justification for why, morally speaking, this is the best template for us to use. Ultimately, while the book shows that we have good reasons to resist the human–robot analogy, it does not address the salient weaknesses of the animal–robot analogy.

We think there is a deeper problem with Darling's account, as well as with other accounts that always strive to keep humans in the loop, even as machines become more autonomous. The problem we see is that the distinction between being responsible and holding responsible is ignored.

This distinction, between being responsible and holding responsible, seems to us both important and underappreciated in resolving the responsibility gap. We're going to explain why and also suggest ways in which the minimalist account of agency can help us work to resolve the responsibility gap. But note, as always, that this is just one account and we're not in the business of proselytizing. You might not agree with our argument, and you might develop strong objections to the account we present. That's awesome. One thing you can consider is, How would you make the account better or stronger or improve it? Critique is an important part of philosophizing, but so is trying to make something a better version than it is. So, read through our account below and try your hand at making it better (even if—or especially if—you object to it). For a demonstration of this point, think about a story of Elizabeth Anscombe, whom we met when we considered action theory. When Anscombe was fourteen years old, she was given a theology book on the existence of god. She devoured it but found it unsatisfying—the arguments given, she thought, were not adequately rigorous. *And so she went and developed a stronger account herself.* This embodies, we think, an important part of doing philosophy well, and we hope you try your hand at improving the work of others.

IV. Problems with the "Human-in-the-Loop" Story

The distinction we draw between being responsible and being held responsible tracks, in part, the one between causal and moral responsibility that has been made since H. L. A Hart drew attention to its importance. Our view is that a whole gamut of causes, agents, and constraints feeds into any happening that is worthy of our taking note (a car crash, a stock market crash, a building being constructed, a harvest) and that it makes sense, and indeed is obligatory, to see that many people and other sorts of worldly systems are responsible for them. This is the first step: discover which systems are involved in the coming to be of this happening and determine the significance of their role. The second step is to lay some of that responsibility at the feet of those who are responsible—to hold them responsible for what they have done. Sometimes, there's nothing more to do in that regard than to note it. Thinking about the space shuttle Challenger disaster, for example, one could simply observe that the heat from the burning hydrogen is responsible for the rapid expansion and cracking of the frozen O-ring in the space shuttle fuel system.

Sometimes, more is required. We need to consider the executives and other non-engineers at NASA who insisted on launching the shuttle, despite engineers telling them what would happen, and then we need to take steps to ensure that things do not go like that again. But in both cases, the first step is to see who and what played a role and what that role is. Until we know whether and what kind of agency was involved in the production of some happening, it is unproductive to try to hold any agents to account. Moreover, there is good reason to think that we need a theory here because we have some evidence to show that people are biased and inconsistent in how they apportion responsibility and blame. Madeleine Clare Elish (2019) refers to the "moral crumble zone," where humans in human–robot interactions are perceived as blameworthy when problems occur, even if the human operator is not completely at fault. A significant problem is that our perception of fault in these situations is often different.

Return to our claim that even a thermostat is an agent. We believe that this way of speaking and thinking about the situation makes sense from a practical point of view as well and not merely from our theoretical commitment to a particular notion of agency. When a room is too hot or too cold, we wonder why. Sometimes, it is because a human has set the thermostat too high or too low, and then we think of that human as responsible for the discomfort we feel. Sometimes, though, the heating or cooling unit is misbehaving, "acting up" we might say, and something must be done about that. Do we blame it in such cases, or praise it when it's working especially well? No, of course not. (Though we have heard machines praised and blamed by their delighted or frustrated users.) But praise and blame are not all there is to responsibility. Instead, identifying who is "in fact responsible" is a matter of identifying the correct locus of agency. "Holding responsible" is then taking whatever remedial steps are required to resolve a claim of loss or apportion a reward, or what have you. But the second steps cannot be taken without the first.

Now consider the raising of children or even the training of pets. One thing that is very clear is that they are agents, even though, in many cases, they cannot be aptly blamed for their misbehaviors. The appropriate response in those cases is further education and deeper inculcation into our social networks. The appropriate response in the case of a misbehaving thermostat is tinkering. It is unclear what the appropriate response is in the case of significantly more autonomous machinery, but it begins by recognizing their agency. Part of any such response will be to return to the designers and builders, but that is part and parcel of holding responsible anyway. We do not merely tinker with the thermostat; we also complain to the manufacturer when the defect can be traced back there. We do not merely educate others' children or pets when they fall afoul of standards; we also complain to their parents and their people when their lack of manners can be traced back there. All of this requires first recognizing the presence of agents and the extent of their embeddedness in our practices of holding responsible. That will be the easier part. More difficult will be to understand how responsibility for various actions diffuses throughout networks that now crucially involve autonomous machinery.

Because we do believe that there really is a problem to resolve about a looming responsibility gap, we part company with those who think these worries about the gap arise from mere speculative thinking or a misunderstanding of the nature of agency. On the other hand, we think that some who have tried to deal with the gap have moved too quickly. This is not a mere ditch that one can jump if only one concentrates and leaps well. Instead, a bridge will be required. Our minimalist account might be seen as a caisson supporting some of the undergirding required to build a bridge across the responsibility gap. Of course, the agents we worry about will be much more sophisticated than thermostats, but if we are not clear on what core features are at the root of agency, then we will have no chance of understanding what is going on in those more sophisticated systems, and we will have no hope of first finding and then holding responsible the right agents. The work here is important but by no means the end of the story; it is a kind of under-laboring and site preparation that is necessary for any finished structure.

V. Wrap Up

Some machines are capable of action. We often fail to recognize this because we are working on the mistaken assumption that action requires beliefs and intentions. And that is at odds with our continuing conviction (and perhaps harping) that a general account of agency should not include any extra mental features (such as beliefs and desires). Once we have adopted the right understanding of agency, we are able to characterize machines as agents when appropriate. Note, however, that the way in which we hold machines responsible for their actions will differ for various reasons; the ascription of agency need not result in one uniform system of praise and blame. Even if both are responsible, a machine (thermostat) that gets the temperature wrong will be regarded differently than a machine (autonomous vehicle) that accidentally kills a person.

As machines become more sophisticated, understanding machine agency and its connection to responsibility becomes even more critical. The introduction of semi- and fully autonomous machines into different spheres of life will disrupt our existing paradigms of ascribing and holding agents responsible.

The next question, and one we must leave to another occasion is, *How do we hold them responsible?*

Your Tasks

Test Your Understanding

1. What are the differences between causal, legal, and moral responsibility? Can you provide an example (similar to the Arjun and Tyler example) that illuminates the distinction?

2. How would you explain the responsibility gap to someone who has never heard of it before?

3. Why does Darling believe that the animal analogy can help us move forward in thinking about how we can assign responsibility for machine actions?

Reflect or Discuss

1. If a self-driving car accidentally kills a person—as has happened in the past—who is responsible and in what sense are they responsible? How would you defend your response to someone who disagrees with you? That is, which reasons would you provide in support of your position?

2. In the responsibility gap literature, there are a lot of different views. We surveyed some of them in this chapter. Return to the brief survey and reflect on each of the views presented. We encourage you to (1) interpret it as charitably as possible; (2) consider whether you think it is compelling, moving, appealing, and so on; and (3) consider potential objections to the view as well as possible ways the author could respond to the objection.

3. How would you close the responsibility gap?

Expand Your Thinking

1. Scholars have carved up the responsibility gap literature in different ways. Recently, Tigard (2020) distinguished between techno-optimists and techno-pessimists. The techno-optimists believe that the responsibility gap can be bridged. Even if they might disagree on how responsibility can or should be found, they typically endorse the view that responsibility can be located somewhere, and they prefer to proceed with the development of AI. The techno-pessimists reject this view, believing that the use of AI leaves us without anyone to hold to account for harms done and that we should scale back production. Note that Tigard characterizes these camps using both descriptive and normative claims. It is plausible, however, to commit to one of the descriptive claims about the responsibility gap (yes,

it can be bridged, or no, it cannot be bridged) without committing to the normative claim (yes, let's keep developing AI, or no, let's halt development). Consider the views presented in this chapter and carve them up according to Tigard's categorization, with the additional separation of normative versus descriptive claims. Then, reflect on where you might fall, and place yourself somewhere in the framework.

2. Ava, Dr. Frankenstein's Creature, and Talos have all killed humans or been involved in the killing of humans. Ava left Caleb to starve and, with the help of Kyoko, killed Nathan; Nathan was keeping her prisoner, but Caleb was not. The Creature killed humans out of anger, revenge, and distress; he felt remorse for some of the killings but continued anyway. Talos killed many who came to Crete by ships, irrespective of why they were arriving. Do you think any of these entities are responsible for what they did? Why or why not?

3. In this chapter, we focus on the responsibility of an agent and what conditions are necessary for attributing responsibility and holding an agent responsible. Hans Jonas's (1984) book *The Imperative of Responsibility: In Search of an Ethics for the Technological Age* provides an example of a different kind of philosophical discussion of responsibility than we have seen in this chapter; Jonas writes that, in the twentieth century, humans incurred the responsibility to choose whether human civilization and diverse, flourishing life on Earth will continue, and points to technological change that has enabled this situation. For Jonas, however, the core question concerning technology and responsibility cannot be answered by thinking about an individual's agency. While Jonas is not thinking about machine responsibility specifically, we invite you to read Jonas's work and consider a different approach to and way into the responsibility question. We also encourage you to think about the other ways the framework for responsibility attribution we have presented in this chapter is inadequate.

Further Reading

Arkin, R. "The Case for Ethical Autonomy in Unmanned Systems." *Journal of Military Ethics* 9, no. 4 (2010): 332–341

Asaro, P. "A Body to Kick, But Still No Soul to Damn: Legal Perspectives on Robotics." In *Robot Ethics: The Ethical and Social Implications of Robotics*, edited by Patrick Lin, Keith Abney, and George Bekey, 169–186. Cambridge, MA: MIT Press, 2011.

Coeckelbergh, M. "Personal Robots, Appearance, and Human Good: A Methodological Reflection on Roboethics." *International Journal of Social Robotics* 1, (2009): 217–221.

Danto, Arthur C. "Causality, Representations, and the Explanation of Actions." *Tulane Studies in Philosophy* 28 (1979): 1–19.

Darling, Kate. *The New Breed.* New York: Henry Holt and Company, 2021.

Elish, Madeleine Clare. "Moral Crumple Zones: Cautionary Tales in Human-Robot Interaction." *Engaging Science, Technology, and Society* 5 (2019): 40–60.

Hanson, F. Allan. "Beyond the Skin Bag: On the Moral Responsibility of Extended Agencies." *Ethics and Information Technology* 11, no. 1 (2009): 91–99.

Hellström, Thomas. "On the Moral Responsibility of Military Robots." *Ethics and Information Technology* 15, no. 12 (2013): 99–107.

Hew, Patrick Chisan. "Artificial moral agents are infeasible with foreseeable technologies." *Ethics and Information Technology* 16, no. 3 (2014): 197–206.

Johnson, Deborah G. "Technology with No Human Responsibility?" *Journal of Business Ethics* 127, no. 4 (2014): 707–715.

Jonas, Hans. *The Imperative of Responsibility.* Chicago: The University of Chicago Press, 1984.

Marino, Dante, and Guglielmo Tamburrini. "Learning Robots and Human Responsibility." *International Review of Information Ethics* 6, no. 12 (2006): 46–51.

Matthias, Andreas. "The Responsibility Gap: Ascribing Responsibility for the Actions of Learning Automata." *Ethics and Information Technology* 6, no. 3 (2004): 175–183.

Nagenborg, Michael, Rafael Capurro, Jutta Weber, and Christoph Pingel. "Ethical Regulations on Robotics in Europe." *AI and Society* 22, no. 3 (2008): 349–366.

Rahwan, Iyad. "Society-in-the-Loop: Programming the Algorithmic Contract." *Ethics and Information Technology* 20, no. 1 (2018): 5–14.

Santoro, Matteo, Dante Marino, and Guglielmo Tamburrini. "Learning Robots Interacting with Humans: From Epistemic Risk to Responsibility." *AI and Society* 22, no. 3 (2008): 301–314.

Sparrow, Robot. "Killer Robots." *Journal of Applied Philosophy* 24, no. 1 (2007): 62–77.

Strawson, P. F. "Freedom and Resentment." *Proceedings of the British Academy* 48 (1962): 1–25.

Tigard, Daniel W. "There is No Techno-Responsibility Gap." *Philosophy and Technology* 34 (2020): 589–607.

Wallach, Wendell and Colin Allen. *Moral Machines.* New York: Oxford University Press, 2010.

11 Agents in the Social World: Moral Status and Relationships

Introduction

The responsibility issue has commanded a great deal of attention in conversations about ever more agential machines. As complicated and important as this question is, it is not the only one. If we develop robust machine agents, we will have to confront many pressing moral and social issues. First, what, if anything, is the moral status of sophisticated machine agents? Second, what kinds of relationships can we and should we have with them? There are, of course, other threads that we can tug on, but these are the two that will anchor our discussion over this exploratory chapter.

We have no settled views on the issues we raise, and will consider a wide range of concepts—not just agency, but also species membership, consciousness, intelligence, and more—to see how useful they are in helping us answer these moral questions.

I. Foundations of Moral Status

Before we can say anything concrete about the possible moral status of machine agents, we need to do a little background work to determine just what we mean by moral status in the first place, and we will also need to sort out what grounds, conveys, or entitles one to that moral status. The difficulty here, as in much of our discussion of agency and mindedness and other concepts covered in other parts of the book, is that we begin the inquiry into the concept of moral status with a limited set of examples of things that fall under that concept. As a consequence, there will be little in the way of clear directions to take when we try to extend that concept to cover, in an anticipatory way, certain kinds of machine agents.

Moral status is, we might say, the thing that I have that makes it unacceptable for you to take away from me things that are mine, including my property, my freedoms, even my life. There are certain circumstances in which it will be acceptable for you to do these things, perhaps, but there is a general prohibition against doing such things to entities who have moral status. Sometimes, people will say things such as "there is a standing prohibition against" various things or that "there is a default presumption" that certain things can or cannot be done to entities with moral status. This can include, for example, a prohibition on arbitrarily killing someone. There are very few philosophers or moral theorists in any discipline who would disagree with the fact that generally humans have that kind of moral status.

When we say that an entity has moral status, what we mean is that the entity has interests for its own sake (and not just for the sake of someone else). Here's a basic example to help us think through this: If I break your smartphone, I may have infringed on your, the owner's, interests—it is, after all, your property. But have I somehow injured the interests of the smartphone? No, because the smartphone does not have interests for its own sake. (Why this is the case is an interesting and important story that we will delve into soon.) Now, consider the case where I accidentally hurt your cat. Just as with the smartphone, I may have injured your interests—the cat is, after all, yours, and perhaps you have to incur the costs of taking the cat to the vet to deal with whatever harm I accidentally caused. But unlike the smartphone, your cat also seems to have interests for its own sake. The cat, a living being with the capacity to experience, does not want to be injured. Because the cat has interests for its own sake, we can say it has some moral status. It's worth noting that some people believe the cat, as a sentient, conscious, and agential living being, is not the kind of entity that humans should be allowed to own as pets. That is, cats ought not to count as property and should not be bought and sold in the way you might buy and sell smartphones or furniture or books.

The concept of moral patiency is also relevant here. A "moral patient" is an entity that is the subject of moral consideration and concern. Moral patients are entities that moral agents (who, as you may remember, have the ability to make choices between right and wrong) should care about and toward which they may have duties. Joanna J. Bryson describes moral patiency as "something a society deems itself responsible for preserving the well-being of" (Bryson 2018, 16). Think of human infants. Human infants, while they will develop into moral agents, are not such as yet—they do not have the capacity to speak, let alone to choose between right or wrong. But we still believe that infants are the subject of moral consideration and concern. In brief, entities who are moral patients or have moral status are worthy of moral consideration.

Where things get contentious is when we look, on the one hand, to find the *justification* or *grounding* of that moral status and, on the other hand, to find out how to *extend* that status beyond the sphere of humankind. (Some would say that it does not make sense to extend moral *status*, for entities outside the sphere of humankind either have or do not have moral status, no matter what we think. If you like, think of extending as increasing our *understanding* of moral status so that we can see whether and which nonhuman entities have it.)

Let's consider a specific case to get a grip on why it can be tough to ground or extend moral status. In the early twenty-first century, many people in the West have become vegetarian or vegan (a dietary practice with a long tradition in many parts of the world). The reasons for this can vary. People may have changed their diet for health reasons or because they are worried about how certain farming practices impact our shared environment. But for some people, the reason for not eating meat is different; it is grounded in their conviction that animals have their own moral status.

If animals have moral status, it is not permissible to eat them (and perhaps not even their by-products such as eggs, milk, or even honey). Our purpose here is not to determine the ethical way to eat (although one of your authors believes that eating a vegan diet is good for the environment, your health, *and* shows animals the moral deference they deserve because of their moral status). Rather, we wonder where this conviction that animals are worthy of moral status comes from. Moreover, we want to explore where such convictions come from *generally* because such convictions are at the heart of why people think of some systems in the world as agents (humans, cephalopods, cows), others as *not* agents (desks and chairs, specks of dust, televisions), and about yet others remain unsure (insects, amoebas, AIs). Our hunch is that there is a connection between moral status, our regard for that status, and our views about what has agency. What needs further exploration is how to determine what exactly that connection entails and what it might tell us about how we should think about machine agents.

We are going to consider a few important voices in the conversation about moral status, and then we will try to draw some conclusions about how this all works. We are not settling this question for you but rather illustrating how to ask it and how to begin to frame answers to it.

You might think that we can simply ground moral status by appealing to species membership. That is, we can say that humans have status because, in virtue of simply *being human*, they are special in a way that nonhumans are not. This is a vastly common view, and indeed underpins many social and legal and cultural practices; we talk, for example, about human rights, which are much more robust than the rights given to other species, ecosystems, and other kinds of entities. This view is not without its criticisms. One of the staunchest critiques, articulated by Peter Singer, is that this view is *speciesist*; it simply takes humans to be intrinsically better than other species without morally justifiable reasons.

Several other approaches to what makes humans different from other entities are tailored to conditions special to humans but do not rely purely on species membership. Generally, these approaches focus on what only humans have or can do. Only humans, it is supposed, have the right kind of interest in their own future, are cognitively sophisticated enough to achieve moral status, and can suffer in the right way.

One of the most prominent grounds for moral status is the possession of certain kinds of cognitive capacities. What exactly this looks like varies from account to account. On some accounts, it's merely the ability to develop sophisticated cognitive capacities; on others, it's the actual possession of sophisticated cognitive capacities. Why does this matter? Because on the first account, we can include infants, toddlers, and children, even if they do not yet have sophisticated cognitive capacities, whereas on the second account, that becomes harder. Many philosophers have also pushed back against grounding moral status in either the possession of sophisticated cognitive capacities or the ability to develop sophisticated cognitive capacities because it is too exclusionary. They propose other grounds for moral status, such as being a member of a cognitively sophisticated species. In "Challenges to Human Equality," Jeff McMahan (2007) posed a provocative account that allows for some human beings to be granted more moral status than others on the basis of cognitive capacities. Many philosophers, including Eva Feder Kittay, have argued strongly against such accounts, and some have argued that they are ableist. These debates in philosophy are controversial because we're dealing with such important and sensitive questions. As you reflect on these different views, keep in mind that these subjects should be considered and discussed in a careful and kind manner.

There are many versions of these approaches and many other similar approaches that we do not consider. Still, they all suffer from the same basic flaw: many humans simply *do not* or *cannot* do these things, or do not or cannot have these things, or, if they can, they demonstrably cannot do so more than or in a way distinct from many nonhuman animals.

In light of this kind of difficulty, some theorists simply stick with the criteria and suggest that since these are the right criteria for moral status then, on the one hand, some humans do not have it, and, on the other, some nonhumans do. We're not at all troubled by the idea that some nonhumans have moral status, but we are troubled by views that exclude the most vulnerable among us. Other theorists find this approach wanting; as an alternative, they propose that we hold humans with full capacities as ideal cases, and then suggest accounts that will bring other humans into the fold, despite their lack of certain capacities that are characteristic of such humans. Instead of *beginning* with being human as the criterion for moral status, they suggest that exemplar humans clearly have moral status by virtue of their exemplary possession of certain cognitive traits, and then *extend* the umbrella of moral status to the rest of the species by a kind of conceptual clustering. Suffice it to say that grounding moral status in a conceptually robust way is far from easy.

II. Machines and Moral Status

The issues get even thornier as we consider sophisticated machines. How should we regard the moral status of machines?

Determining whether an entity has moral status is crucial because it helps us know how the entity should be treated and what duties are owed to it. Consider, for example, that debates regarding the treatment of animals are shaped by questions of moral status. How we answer questions such as "Should we burn off the beaks of chickens?" and "Is it ethical to kill animals who destroy human crops?" depends on the kind and degree of moral status we take the animal in question to have.

In the previous section, we focused on using cognitive capacities to ground moral status. If (1) this view of moral status is right *and* (2) it is possible to create machines with sophisticated cognitive capacities, then it is easy to see why machines could be said to have moral status. But if you disagree with grounding moral status in cognitive capacities or you do not think it's possible to create machines with the right kind of sophisticated cognitive capacities, then it would not make sense to view machines as having moral status *on this basis*. But perhaps machines could have moral status on another basis.

You might think, for example, that the capacity to suffer is the right kind of grounding for moral status. If machines had the capacity to suffer, you might be even more inclined to think that they have interests for their own sake. Which features would a machine need to have in order for us to be convinced that it has moral status?

John Basl and Henry Shevlin, respectively, believe that consciousness is important for grounding moral status. For Basl, "In the near future, no matter how complex the machines we develop, so long as they are not conscious, we may, as far as concerns the artifact itself, do with them largely as we please. However, things change once we develop, or think we are close to developing, artificial consciousness. Once artificial consciousnesses exist that have the capacity for attitudes they have psychological interests that ground their status as moral patients" (Basl 2014, 95). Shevlin is concerned with the epistemic problem Basl points to: How would we know that machines had consciousness? Shevlin goes on to argue that a machine should be considered a psychological moral patient to the extent that it possesses cognitive mechanisms shared with other beings such as nonhuman animals whom we also consider to be psychological moral patients.

Basl and Shevlin both explore moral status in the context of moral patiency. They are not asking what a machine needs to do in order for us to be convinced that the machine can make choices between right and wrong. Rather, they are concerned with when and how a machine might be considered a moral patient. Why does this matter? Because you do not need to be a moral agent to have moral status. Irrespective of whether machines have moral agency, they can still have moral patiency—and be the appropriate subjects of consideration.

But could agency itself be a grounding for moral status? Put differently, even if the more common groundings for moral status—cognitive capacities, capacity to feel pain, and so on—did not apply to machines, could we ground their moral status in their agency?

When an entity has agency, we respond to and regard it differently. Think back to our fictional stories. In Ava's case, because of her sophistication and first partial and then full mimicry of human physiology, it's easy for the viewer to imagine the emotional attachment that Caleb forms with her and to sympathize with Caleb's thought that she is owed his help in escaping simply because she's a person in trouble and he's on the scene. All of that arises because the maker of the film takes advantage of the fact that many of our moral judgments are, in fact, moral sensations and that they arise from other sources than a conceptual analysis of the situation. Caleb thinks he's making a dispassionate appraisal of Ava's state of consciousness until he finds that he's been manipulated by Nathan to react to Ava sexually, and even so he continues to think that she can make moral demands even after finding this out. The audience is largely in the same position.

Why does it not seem absurd to think that Ava might make moral demands on us? Is it simply that, to just the right extent, she *looks* like us or *sounds* like us? How similar does a machine have to be for it to make moral demands on us? What if there are limits to those similarities? And is it perceptions of agency, or intelligence, or sentience, or consciousnesses, or something else that is doing the important work?

This is more than a hypothetical question. In the summer of 2022, a Google engineer, Blake Lemoine, raised the alarm that Google's chatbot, LaMDA, was sentient. In a *Washington Post* interview, Blake said, "I know a person when I talk to *it* . . . it doesn't matter whether they have a brain made of meat in their heads. Or if they have a billion lines of code. I talk to *them*. And I hear what they have to say, and that is how I decide what is and isn't a person" (Tiku, 2022, emphasis added). He responded to LaMDA because of the way it spoke and, as he claimed, communicated with him.

What about systems that are less sexy, and less conversational, and less whatever it is that makes us respond to them emotionally? Would we really want a theory of morality that relied on considerations such as the look, sound, or even smell of something—considerations that are tied directly to our evolutionary heritage and the ways we are built to respond to each other?

Perhaps, though, that is not what is going on here after all. Perhaps consciousness is the key. An easy suggestion in this case is that the reason we treat Ava as (more or less) human is that she passed Caleb's version of the Turing test—a test that Nathan asked him to use to determine whether she is conscious. Most theorists agree, however, that the Turing test supplies neither necessary nor sufficient conditions for assessing the consciousness of those who take it. While consciousness may be important, then, we are no closer than ever to being able to test for it rather than simply seeing it. There are other tests for AGI, but those only test for certain capacities to perform certain tasks, and as we saw above, it is a pretty dicey matter to try to use various capacities, in fact, to make appraisals of moral worth. Where, then, does that leave us?

Many of the possibilities here rule out of court machines as moral agents or patients. Or they are so epistemically fraught or conceptually indeterminate that they cannot be used as a framework for understanding whether machines ever could have moral status. That is perhaps just the best that can be done at the moment. But there is another possibility, one that we explore in the next section. That possibility is that it is the capacity to enter into morally salient relationships or into networks of mutual dependency that provides a *general* grounding for moral status. Agency, of the sort we have been discussing, might make something of the right sort to be *considered* for moral status, while the entity's actual capacities to enter into such relationships or networks allows it to *have* that status.

III.   What's So Special about Relationships?

Boomer, a robot used to detect bombs, "died" in 2013. At least, that is how it seemed to many of the American soldiers who gave Boomer a funeral.

There's a deep irony to holding a funeral for a robot in the middle of a war when, nameless and faceless, many civilians who died—humans with subjective experiences and feelings and family histories—went without one. Why would American soldiers throw a funeral for bits and pieces of machinery? The most charitable and plausible reading, on our view, is that the American soldiers had a relationship with Boomer, and it is the specialness of this relationship that made holding a funeral for a robot fitting. Perhaps the bizarreness never even occurred to them, given the range of experiences they had had with Boomer on the battlefield. This example, you might be interested to note, is not such an outlier. Colin Angle, for example, has said that many soldiers send requests to have their PackBot (another kind of robot used in the battlefield) fixed (Wallach and Allen 2010).

Humans have relationships with all sorts of entities and things in the world, although, as we saw, there are disagreements about what constitutes a relationship. Relationships deeply matter to us, and the kinds of relationships we have also deeply matter to others around us who are impacted by our relationships. As Wendall Wallach and Colin Allen write in *Moral Machines*, "like it or not, existing robots are not just passive conduits of ethical rules but themselves interact with other agents in an existing moral ecology. The soldier's concern for his bomb-sniffing robot introduces new ethical possibilities, for example, how he would rank the survival of the robot against that of, say, a dog" (Wallach and Allen 2010, 61). To extend Wallach and Allen's point, in the context of human–machine relationships, there are interesting questions to ask not only about how these relationships affect individuals but also about how they affect everyone and everything else in the existing moral ecology.

There are many more examples that we can keep exploring that help us glean something about relationships. But before jumping into any more, let us cover the basics. What is a relationship? Here is a minimalist definition: when two things are said to be in a relationship, all that is meant is that they stand in a kind of *relation* to each other. Note, here, that a relation is a general concept, and it encompasses more than human relations with other humans—that is a special kind, which we will return to soon. There are different kinds of relations; things may be temporally related, spatially related, causally related, and so on. There are also limitations to what kinds of relations we can have with one another. I cannot, as one example, be your biological parent—you are already born, and there are only, biologically speaking, two people who can stand in relation to you as biological parents.

As with most things in philosophy, there is ample disagreement about what constitutes a relationship. Many philosophers would look at the minimal conceptualization articulated above and say, "Yes, *and . . .*" For some philosophers, a relationship needs to be at least minimally *reciprocal* to count as a relationship. This does not mean that it needs to be *symmetrical*. Relationships such as that of a parent and child, at least in the early stages of life, are far from symmetrical. But you might argue there is still reciprocity: parents give love, care, and security, and children give attitude, naughtiness, and other grief that makes you regret procreating. This is a (half-)joke, of course, but there's a deeper point: children are the kinds of entities that are *capable of reciprocating* in some way, even if they do not in fact reciprocate in the same way or in the way we might want. Indeed, we navigate relationships and decide which ones we want to stay in—with a friend, with a partner, with a boss—at least in part depending on what we get from the other person.

If you think reciprocity is needed for two entities to be in a relationship with one another, then there are additional interesting questions you could explore about what is needed for reciprocity or for being the kind of entity that can reciprocate. What about plants, for example? They seem largely passive, and while they do give off oxygen, among other activities, that seems to have very little to do with our own comportment toward them. Still, they grow and give pleasure, and maybe that is all that is needed.

Consider now another story involving relationships between humans and inanimate objects. In the movie *Castaway*, Tom Hanks's character, stranded on an island without companionship, forges a relationship with an inanimate object, a volleyball he named "Wilson." He speaks to the ball, paints a face on it, and generally centers his life around it in the way one would with an important human in one's life. But a volleyball such as Wilson is not the kind of entity that can reciprocate. This even goes far beyond the case of relationships with plants. At best, Hanks anthropomorphized the volleyball and projected onto it feelings and intentions and so on.

Reading this, you might well conclude that the capacity to reciprocate is not a necessary feature for one to stand in a relationship with another. We have relationships with all sorts of objects in the world—our social media, alcohol, our work—and making sense of a general account of relationships does not require reciprocity or the capacity to reciprocate. We do not know what the best account is. But we do think that irrespective of whether something counts, conceptually speaking, as a relationship in the more demanding sense has little bearing on the kinds of interesting philosophical questions we can ask about the relationship.

Relationships are an integral part of our social and moral and cultural universe. Many of the roles and responsibilities we have are anchored in a specific kind of relationship: the role and responsibilities I have toward my mum as a daughter, for example, are different from those that my friend has toward my mum as the friend of a daughter. We saw this when we discussed different accounts of responsibility and how a doctor might have a different set of responsibilities in virtue of her role. Another way to articulate this is to say our roles are, in part, a reflection of our relationships in a particular context with others. As a doctor, I stand in a doctor–patient relationship with my patients. But when I, the doctor, go home, take off my scrubs, and make dinner with my wife, I stand in a wife–wife relationship with her, not in a doctor–patient relationship. As you can see, roles, responsibilities, and relationships are deeply entangled.

For some philosophers, relationships ground other kinds of morally significant concepts, such as moral status. For Kittay, moral status is grounded in social relations. She writes:

> By a "social relation" I mean a place in a matrix of relationships embedded in social practices through which the relations acquire meanings. It is by virtue of the meanings that the relationships acquire in social practices that duties are delineated, ways we enter and exit relationships are determined, emotional responses are deemed appropriate, and so forth. A social relation in this sense need not be dependent on ongoing interpersonal relationships between conscious individuals [e.g., it can involve the incapacitated or even the dead] . . . Identities that we acquire are ones in which social relations play a constitutive role, *conferring moral status and moral duties.* (Kittay 2005, 111)

Similarly, in "Person-Rearing Relationships as a Key to Higher Moral Status," Agnieszka Jaworska and Julie Tannenbaum aim to develop an account of moral status that can accommodate the intuition that humans, including children (who are not yet rational), should be granted higher moral status than most animals. They argue that the difference between a child and a dog lies with the child's capacity to engage in person-rearing relationships with the goal of self-standing personhood, which grounds higher moral status. What should be clear by now is that relationships ground all sorts of moral commitments and duties and, for some philosophers, even moral status.

Of course, many of these philosophers did not have machines in mind when discussing moral status and relationships; we note their work because it shows how relationships have been used to anchor claims to moral status. But what *about* machines?

Recall Blake Lemoine's story. Many of us will think that he was wrong to attribute intelligence to LaMDA because its language use was ersatz rather than genuine. Even so, Lemoine's conviction that he was in a relationship with LaMDA, a person on his view, prompted him to try and secure rights and privileges for it. At the moment, few of us are able to see LaMDA as a rights-deserving entity. Notice that this may have broader effects if enough people come to share Lemoine's moral conviction; perhaps our assessment of its moral status will change in the way that it did for animals in the twentieth century.

IV. Relationships with Machines

For some observers, the story of Boomer and the American soldiers, or Lemoine and LaMDA, had a striking parallel to that of Tom Hanks and the volleyball, Wilson, in *Castaway*. Hanks's character is stuck on an island with no other companionship, and he enters into a relationship with an inanimate object for very understandable reasons: to avoid loneliness, to stay sane, to have a friend. In a poignant scene at the end of the film, Hanks loses Wilson to the ocean. The pain Hanks feels is presented as excruciating, and the audience vicariously experiences the sorrow and terror of watching a loved one drown. That is what the volleyball is for Hanks: a loved one that drowns.

But notice here that Hanks's case is tragic precisely because he is isolated on a desert island with no human companionship.

We might claim, reasonably, that the kind of relationship Hanks has with the ball is not one we should promote or foster or encourage, even if it is one that, when it occurs in rare and sad circumstances, we understand. This opens up interesting questions for us: What kinds of relationships should we promote? Could there be relationships with nonhumans that are bad for humans to have? Are unilateral affective relationships good or bad for humans? Should AI occupy care roles?

These are complex questions, and they are difficult to answer in no small part because many of us disagree on whether the AI or machines we have in front of us actually are capable of agency. Many researchers who focus on human–robot relationships work from the reasonable assumption that robots are not minded, conscious, or intelligent like humans are. So, they ask questions about human perceptions of agency and mindedness and how that might be good or bad for us.

Our relationships and, more generally, the attachments we form, have to do with our (the attachees) perceptions of the object we're attaching to. We have a different relationship with a rock than we do with a cat in no small part because we perceive the cat to be agential, sentient, and conscious. It turns out that our threshold for seeing something to be agential is very low. Recall the 1940s study by Heider and Simmel that we discussed earlier in the book, where participants attributed agency and purpose to black-and-white geometric shapes moving around on a screen. As we saw, this is at least partly explained by our evolutionary history: because we look out for predators, our brains perk up when something agential is in front of us.

There is a lot of research that suggests that not only do we see agency but also that we have higher expectations of machines that appear agential. A research project on Microsoft's Clippy explained why it was so hated by users: users perceived Clippy to be an agent instead of a tool and were annoyed by it when it did not act according to social conventions (Nass 2010). Masahiro Mori's concept of the uncanny valley—the dissonance between what appears to be human but fails to meet human expectations—is worth keeping in mind.

As Kate Darling helpfully explains,

> People think a robot that is present in a room with them is more enjoyable than the same robot on a screen and will follow its gaze, mimic its behavior, and be more willing to take the physical robot's advice. We speak more to embodied robots, smile more, and are more likely to want to interact with them again. People are more willing to obey orders from a physical robot than a computer. When left alone in a room and given the opportunity to cheat on a game, people cheat less when a robot is with them. (Darling 2021, 100)

What does this tell us? It at least tells us that some of our responses to machines depend on our perception of their agency and not on whether they are in fact agents.

V. Looking Forward

This book has been focused on whether machines could be agents, how to make that happen if so, and then what sorts of agents they might be. We had our own view about agency, and while we tried to explain what is good about that view, we also explored what it would be like to have alternate, competing, or complementary views to our own. Here, though, on questions about what might give machines moral status and what the connection is between moral status and genuine relationships between humans and machines, we do not have settled views. In fact, what we will be doing in this final chapter is getting oriented toward how to frame the right questions that one might need to address in order to develop a useful theory that ties agency, moral status, and relationships together so that we can understand what it may be like as machines become embedded in our social economy. We do not have that, and we hope this is a project you consider taking up.

One useful concept to bring to this chapter is the concept of "enabling condition." Enabling conditions are not enough to make some one thing count as some other thing, but they can help to make them candidates. For example, being quick with mathematics and physics formulae does not make one an engineer, but it can certainly help enable one to acquire the other necessary skills and attitudes to be one. Our initial thought is that agency is certainly not sufficient to give some entity moral status, or to make it an appropriate target for a meaningful relationship—but it may be an important enabling condition. See for yourself.

It's time to rethink this whole line of inquiry from the beginning. Indeed, that rethinking is, in some respects, the point of the book itself. We have been speaking throughout this book about what it would be like for certain kinds of machines of our devising to be the sort of thing we would all see as being responsible for actions. We talked a lot about agency, and we talked a lot about what it would take to implement more than minimalist agency.

Many philosophers have begun by analyzing humans as agents and use that to determine what capacities various machines do or do not, may or may not have in the future. This is a standard way of doing things that strikes us as in need of an inversion. We do not expect to learn anything very useful about agency *in general* by further introspection into the nature of our own inner states and into the way that agency is implemented by means of our own special representations of belief, and desire, and so forth. Instead, we think, we will finally learn more about agency in general, and in machines more specifically, when we de-center the human.

We don't know where things will go with AI. When we began writing this book, there was much more buzz about self-driving cars than there currently is, and many companies have abandoned projects on embodiment and robots. It turns out the hardware problems are harder to solve.

Your Tasks

Test Your Understanding
1. What is the difference between moral agency and moral patiency?
2. Is reciprocity needed for something to constitute a relationship?
3. Name three possible groundings for moral status discussed in this chapter.

Reflect or Discuss
1. What do we want from a theory of moral status? That is, what does talking about moral status give us?

2. Reflect on the distinction between AI, a robot, and a computer. We can imagine a robot that is not AI, or AI that is not embodied in a robot, or a computer that is not a robot or AI. In what ways might the kind of entity and object shape or, more strongly, determine the kinds of relationships we can have with it?

3. Reflect on the relationship between moral status, responsibility, and agency. What are the connections you see? How might the concepts fit together in one theory?

Expand Your Thinking

1. Turn back to the fictional entities we discussed in the chapter "Myths of Machine Agents". Using the concepts we have explored in this chapter, how would you evaluate the moral status of each of the fictional entities? What role are relationships playing in the stories? Why might the humans in the stories be seeing agency when the entities are not in fact agential?

2. In the 2013 movie *Her*, some speculative ideas about the future of AI are explored. Samantha is an intelligent computer system whose exact nature is not fully spelled out, although we know that she is disembodied. She is, however, capable of expressing empathy and other human emotions, and she enters into a romantic relationship with Theodore. At the end of the film, Samantha and other AI systems around the world seem to merge into a hypermind and then go . . . somewhere . . . a kind of metaphorical place that humans cannot really understand. In contrast with pessimistic imaginings of the future development of artificial agents, *Her* imagines a future where super intelligences, far from being malicious, are actually capable of loving humans and respecting their autonomy, even as they become more and more intellectually advanced. This kind of vision, of smarter and smarter agents who, in part as a result of their increasing intelligence, are also capable of human emotions, is clearly of the speculative sort. We can surely wonder what it would be like to be in interaction with such agents, even as it is not entirely possible to imagine the exact parameters of such systems fully. Samantha's relationship with Theodore confronts us sharply with questions about what constitutes and sustains our relationships with other humans, as it asks us to imagine relationships that are missing some of the things that we might normally take to be fundamental to relationships between

friends and lovers: physicality generally, the registering of emotions in body language and facial expression, the quotidian nature of sharing space and various objects inside that space, and so forth. Reflect on this story and others. Here are some questions to aid your reflection. Do you think, were it possible, humans should be in relationships with advanced machines? Should we allow marriage between a machine and a person? What about dating? What about sex?

3. Turn back to the questions you wrote for yourself at the end of the "Orientation" chapter. Try to answer the questions now, and then write down three new questions you have for yourself after reading this book. Then, go out and, using the philosophical skills you've developed, try your hand at answering them.

Further Reading

Basl, John. "Machines as Moral Patients We Shouldn't Care About (Yet): The Interests and Welfare of Current Machines." *Philosophy and Technology* 27 (2014): 79–96.

Bryson, Joanna J. "Patiency is not a Virtue: The Design of Intelligent Systems and the System of Ethics." *Ethics and Information Technology* 20, (2018) 15–26.

Buchanan, Allen. "Moral Status and Human Enhancement." *Philosophy and Public Affairs* 37, no. 4 (2009): 346–381.

Clarke, Steve, Hazem Zohny, and Julian Savulescu, eds. *Rethinking Moral Status.* Oxford: Oxford University Press, 2021.

DeGrazia, David. "Moral Status as a Matter of Degree?" *The Southern Journal of Philosophy* 46, no. 2 (2008): 181–198.

Jaworska, Agnieszka and Julie Tannenbaum. "Person-Rearing Relationships as a Key to Higher Moral Status." *Ethics* 124, no. 1 (2014): 242–271

Kittay, Eva Feder. "At the Margins of Moral Personhood." *Ethics* 116, no. 1 (2005): 100–131.

McMahan, Jeff. "Challenges to Human Equality." *The Journal of Ethics* 12, no. 1 (2007): 81–104.

Nass, Clifford Ivar. *The Man Who Lied to His Laptop.* New York: Current, 2010.

Schneider, Susan. *Artificial You.* Princeton: Princeton University Press, 2019.

Shevlin, Henry. "How Could We Know When a Robot Was a Moral Patient?" *Cambridge Quarterly of Healthcare* 30, no. 3 (2021): 459–471.

Singer, Peter. *Animal Liberation: A New Ethics for Our Treatment of Animals*. New York: HarperCollins, 1975.

Tiku, Nitasha. "The Google Engineer Who Thinks the Company's AI Has Come to Life" *The Washington Post*, June 11, 2022. https://www.washingtonpost.com/technology/2022/06/11/google-ai-lamda-blake-lemoine/

Warren, Mary Anne. *Moral Status: Obligations to Persons and Other Living Things*. Oxford: Clarendon Press, 1997.

Index